Mitochondria

Integrated Themes in Biology

Consulting Editor: I. D. J. Phillips, University of Exeter

Hall and Baker: Cell Membranes and Ion Transport
Pitt: Lysosomes and Cell Function
Whittaker and Danks: Mitochondria

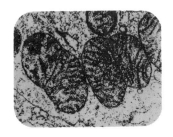

Mitochondria: structure, function and assembly

Peter A. Whittaker and Susan M. Danks

School of Biological Sciences
University of Sussex

Longman LONDON and NEW YORK

Longman Group Limited London

*Associated companies, branches and representatives
throughout the world*

*Published in the United States of America
by Longman Inc., New York*

First published 1978

Library of Congress Cataloging in Publication Data

Whittaker, Peter Anthony, 1939—
    Mitochondria: structure, function, and assembly.

    (Integrated themes in biology)
    Includes bibliographies and index.
    1. Mitochondria.   I. Danks, Susan M., joint
author.   II. Title;
QH603.M5W48   574.8'734   77-13966
ISBN 0-582-44382-2

Printed in Great Britain by
Whitstable Litho Ltd., Whitstable, Kent

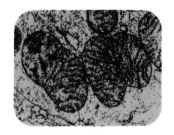

Preface

It is our feeling that there is a gap in the range of literature available for teaching undergraduate courses on mitochondria. Books tend to be either aimed at 'A' level or first-year undergraduate students or postgraduate workers involved in research into mitochondria. Therefore, important concepts involved in the study of mitochondria are explained either at a very superficial level or else too technically. We hope our book will fill this gap. In addition, we have tried to draw together all aspects of mitochondria and present them in a coherent fashion such that the structure, function and assembly of the organelle can be clearly interrelated.

We have not included references in the text, but have given a reference list at the end of each chapter with at least one reference from each topic in the chapter. This could serve as the basis for reading by advanced third-year students or even postgraduate workers. We have attempted to adopt a practical approach to the subject indicating and summarising where possible some of the more important research techniques which have been employed.

We would like to point out a particular problem which we encountered in the case of the words 'cytoplasm' or 'cytoplasmic' in connection with mitochondrial biogenesis. Some authors have used the terms in the context of cytoplasmic inheritance — meaning, in this case, a probable location of a gene on mitochondrial DNA and on the other hand 'cytoplasmic ribosomes', meaning extramitochondrial ribosomes. The authors understand cytoplasm to refer to all the cell except the nucleus, i.e. it includes the mitochondria. We understand cytosol to be the supernatant which remains after nuclei, mitochondria, endoplasmic reticulum and ribosomes have been removed from the cell homogenate by centrifugation. We have tried to use the appropriate terms.

We would like to thank Dr John D. McGivan and Barbara M. Allmark for reading and making helpful comments on various parts of the manuscript.

Contents

	Preface	v
	Acknowledgements	x
	Abbreviations	xi

1	Introduction	1
1.1	Distribution and location of mitochondria	1
1.2	Structure and ultrastructure	3
1.3	Functional organisation of mitochondria	9
	1.3.1 Preparation of mitochondria	9
	1.3.2 Localisation of mitochondrial components	13
1.4	ATP and biological work	17

2	Mitochondrial metabolism	22
2.1	Introduction	22
2.2	The tricarboxylic acid cycle	23
2.3	The central role of the TCA cycle	25
2.4	Anaplerotic reactions	29
2.5	Structural organisation of the TCA cycle	32
2.6	The urea cycle	36
2.7	β-Oxidation of fatty acids	38
2.8	Regulation of mitochondrial metabolism	41
2.9	Other metabolic reactions occurring in mitochondria	45

3	Oxidative phosphorylation	47
3.1	Introduction	47

3.2   Electron carriers   48

    3.2.1   Pyridine-nucleotide-linked dehydrogenases   48
    3.2.2   Flavoproteins   50
    3.2.3   Iron-sulphur proteins   50
    3.2.4   Ubiquinone   51
    3.2.5   Cytochromes   54
    3.2.6   Copper   59

3.3   Oxidation-reduction (redox) potentials   60
3.4   Structural organisation of the electron transport system   61
3.5   Inhibition of the respiratory chain   65
3.6   Measurement of oxidative phosphorylation   68
3.7   The coupling of ATP synthesis to electron flow   70

    3.7.1   The chemical coupling theory   71
    3.7.2   The chemiosmotic coupling theory   73
    3.7.3   The conformational coupling theory   79
    3.7.4   Discussion of the coupling mechanism   80

3.8   The adenosine triphosphatase complex   81

    3.8.1   The exchange reactions   83
    3.8.2   The isolated oligomycin-sensitive ATPase complex   84
    3.8.3   The structure of the ATPase complex   86

3.9   Reversed electron flow   89
3.10   NAD(P)$^+$ transhydrogenase   88

4   Transport of ions across the mitochondrial membrane   91

4.1   Introduction   91
4.2   Methods of determining mitochondrial permeability   92
4.3   Types of transporter   95
4.4   Anion transporters   98
4.5   The biological importance of the transporters   102
4.6   The transfer of reducing equivalents into the matrix   102
4.7   Electrogenic transporters   104

    4.7.1   Glutamate/aspartate exchange   104
    4.7.2   Ornithine/citrulline transport   105
    4.7.3   Glutamine/glutamate exchange   105
    4.7.4   Adenine nucleotide exchange   105

4.8   Carnitine — the transport of fatty acyl CoA   106
4.9   Transport of cations   107

    4.9.1   Monovalent cations   107
    4.9.2   Divalent cations   108

4.10   Isolation of proteins associated with transport   109

5	**The assembly of mitochondria**	112
5.1	Introduction	112
5.2	Cellular origin of mitochondria	113
5.3	Mitochondrial nucleic acids and protein synthesis	115
5.4	Mitochondrial mutants	122
5.5	Mitochondrial assembly	127
5.6	Mitochondrial genetics	132
5.7	Evolutionary origin of mitochondria	135

Index	139

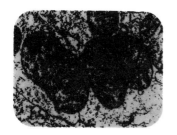

Acknowledgements

We are grateful to the following publishers and respective authors for permission to reproduce copyright material:

Academic Press Inc. (London) Limited for Fig. 5.3A by M. M. K. Nass *et al.* from *Journal of Molecular Biology* Vol. 54 (1970); Academic Press Inc. New York for Fig. 4 by N. Bjorkman *et al.* from *Experimental Cell Research* Vol. 27 (1962); American Association for the Advancement of Science for Fig. 2 by H. Hoffman and C. J. Avers from *Science* Vol. 181 (1973). Copyright © 1973 by the American Association for the Advancement of Science; Cold Spring Harbor Laboratory for part of Fig. 1a by D. R. Wolstenholme *et al.* from *Cold Spring Harbor Symposium on Quantative Biology* Vol. 38 (1973); Elsevier/North-Holland Biomedical Press for Fig. 4a by Wrigglesworth *et al.*, Vol. 205, Fig. 5c by Hollenberg *et al.*, Vol. 209 and adaptation of Fig. 2 by A. E. Senior, Vol. 301 from *Biochemica et Biophysica Acta;* Federation of American Societies for Experimental Biology for Figs. 2 and 3 by E. Racker, reprinted from *Federation Proceedings* Vol. 26 (1967); The Royal Society for Fig. 5A.B by Greville *et al.* from *Proceedings of the Royal Society* (B) 161, (1965); Springer-Verlag for Fig. 7 by G. F. Leedale *et al.* from *Arch. für Mikrobiologie* Vol. 50 (1965); The Rockefeller University Press for Fig. 7 by W. J. Larsen from *Journal of Cell Biology* Vol. 47 (1970).

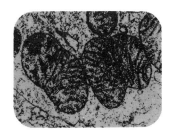

Abbreviations

AMP	Adenosine-5'-monophosphate
ANS	1-Anilino-8-naphthalene sulphonic acid
ADP	Adenosine-5'-diphosphate
ATP	Adenosine-5'-triphosphate
ATPase	Adenosine-5'-triphosphatase (ATP phosphohydrolase)
BR	Bacteriorhodopsin
CoA	Coenzyme A
CTP	Cytidine-5'-triphosphate
Cyt	Cytochrome
DABS	2,5-Diaminobenzenesulphonic acid
DCCD	N,N'-Dicyclohexylcarbodiimide
DNA	Deoxyribonucleic acid
DNP	2,4-Dinitrophenol
EC	Enzyme Commission
ESR	Electron spin resonance
ETF	Electron transferring flavoprotein
$FAD/FADH_2$	Flavin adenine dinucleotide (oxidised/reduced)
FCCP	p-trifluoromethoxycarbonylcyanide phenylhydrazone
$FMN/FMNH_2$	Flavin mononucleotide (oxidised/reduced)
GDP	Guanosine-5'-diphosphate
GTP	Guanosine-5'-triphosphate
ITP	Inosine-5'-triphosphate
K_d	Dissociation constant
K_i	Inhibitor constant
LA	Lipoic acid
MI	Mitochondrial inhibitor
M_r	Relative molecular mass ($\hateq$ molecular weight)
mRNA	Messenger ribonucleic acid
$NAD^+/NADH$	Nicotinamide adenine dinucleotide (oxidised/reduced)
$NADP^+/NADPH$	Nicotinamide adenine dinucleotide phosphate (oxidised/ reduced)
OSCP	Oligomycin sensitivity conferring protein
PFK	Phosphofructokinase

Pi	Inorganic phosphate
p.m.f.	Proton motive force
poly A	Polyadenylic acid
PPi	Pyrophosphate
RCR	Respiratory control ratio
RNA	Ribonucleic acid
SDS	Sodium dodecyl sulphate
TMPD	N,N,N',N'-Tetramethyl-p-phenylenediamine
TPP	Thiamine pyrophosphate
tRNA	Transfer ribonucleic acid
UTP	Uridine-5'-triphosphate
$UQ/UQH_2(Q/QH_2)$	Ubiquinone (oxidised/reduced)

Chapter 1

Introduction

1.1 Distribution and location of mitochondria

Living organisms can be divided into two major groups. The more advanced group comprises the eukaryotes, whose distinguishing feature is the presence of a true nucleus containing chromosomes and bounded by a nuclear membrane. The other group of organisms comprises the prokaryotes, which do not possess these features but characteristically have a nucleoid consisting of a tightly packed skein of DNA which lacks a limiting nuclear membrane. Bacteria and blue-green algae are prokaryotes – all other organisms are eukaryotes. Apart from the differences in the organisation of the genetic material there are a number of other structural and bio-chemical features characteristically different in the two groups. One of the most significant of these is the possession by eukaryotes of mitochondria. As we shall see the major function of mitochondria (although by no means the only one) is the synthesis of ATP from ADP and inorganic phosphate (Pi) by the process of oxidative phosphorylation. Prokaryotes, which do not contain mitochondria, carry out oxidative phosphorylation either on the outer limiting membrane of the organism or possibly on invaginations of this membrane (mesosomes).

The evolution of mitochondria, which must have multiplied the cell's capacity for ATP synthesis, was probably one of the crucial factors which permitted the evolution of larger cells and ultimately the range of sophisti-cated multicellular organisms which exists today. Clearly an understanding of the structure and operation of the mitochondria and their interaction with the whole cell must be essential to the appreciation of the success of eukaryotes.

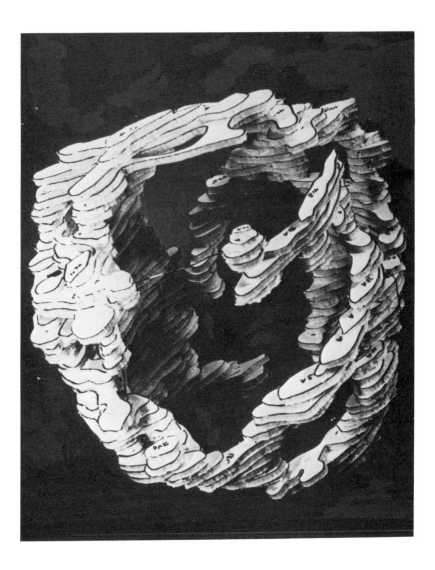

Fig. 1.1 Model of yeast cell mitochondria. The models were constructed by mapping the mitochondrial profiles in serial sections through yeast cells, and assembling them to give a three-dimensional map of the yeast cell mitochondria. This model represents the mitochondrion of a non-budding stationary phase cell. (Adapted from photograph courtesy of Dr C. Avers.)

The numbers of mitochondria in eukaryotic cells vary enormously. A small number of species contain no mitochondria at all. These are invariably simple unicellular species which have probably lost their mitochondria by degenerative evolution. Special cases are the *petite* colony mutants of yeast which lack functional mitochondria (see Chapter 5) and mammalian red blood cells whose mitochondria degenerate during the course of their development. Most of these mitochondria-less cells obtain their energy supply of ATP by glycolytic fermentation, although it is possible that some unicellular algae or protozoa may obtain ATP from an endosymbiotic bacterial species harboured in their cytoplasm. At the other end of the scale are the ova of some sea urchins which can contain many thousands of mitochondria. A number of species, such as the flagellate alga *Chromulina pusilla*, contain only a single mitochondrion which divides into two during the course of cell division. The crithidial (non-reproductive) stage of trypanosomes also contains just a single large mitochondrion. Recently it was suggested that serial sections of the yeast *Saccharomyces cerevisiae* showed that it contained only a single highly branched mitochondrion forming a network around the periphery of the cytoplasm (Fig. 1.1). It now appears that this network can be more or less fragmented into a smaller number of mitochondrial structures depending on the growth conditions of the cell. Rat liver cells, which have been much used in isolating mitochondria for research purposes, contain something of the order of 1 000 mitochondria.

It is important to remember that the individual distinct mitochondria seen in the fixed cells may be merely a static picture of a dynamic situation. Phase contrast films of living cells show that mitochondria can fuse and break up, so that it may be that mitochondria can exist as a temporal continuum within the cell. This is probably not always the case, however, as some mitochondria have a more or less fixed cellular location. As might be expected mitochondria are often located close to the place where ATP is required. This is particularly noticeable in insect flight muscle and other muscle cells where the mitochondria are sandwiched between the myofibrils (Fig. 1.2) and spermatozoa where they are helically wound around the base of the flagellum.

1.2   Structure and ultrastructure

It is not possible, in a book of this size, to detail the enormous variety of mitochondrial structures which exist. This section concentrates on the unifying features of mitochondria with a brief mention of important variations. This is not to say that the variety of structures is not important, almost certainly the differences have evolved to permit the organelles to carry out their similar functions in their different cellular environments.

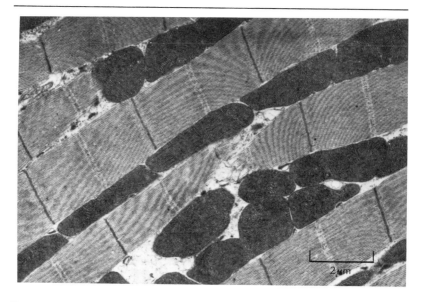

Fig. 1.2   Location of mitochondria from *Calliphora erythrocephala* (blowfly) flight muscle mitochondria. Glutaraldehyde/osmium tetroxide fixed; uranyl acetate/lead citrate stained. (Courtesy Dr M. Tribe and Mrs S. Webb.)

The overall shape of mitochondria is variable. In many cases they have a fairly regular shape varying from spherical to cylindrical. Spherical mitochondria vary from 0.5–5 μm in diameter whereas cylindrical ones tend to be 0.2 μm or more in diameter and up to 10 μm in length. Individual species or cell types may exhibit rather specialised shapes, such as cup-shapes, annuli or discs. Such shapes greatly increase the surface area to volume ratio of the mitochondrion and could increase the potentiality for exchange of metabolites between the cytosol and the mitochondrion.

Fundamentally a mitochondrion consists of a double membrane structure, and these two membranes define two spaces, the intermembrane space and the matrix. Figure 1.3(*a*) and (*c*) show electron micrographs of rat liver and rat anterior pituitary somatotroph (RAPS) mitochondria respectively. The rat liver mitochondrial electron micrograph was prepared using conventional techniques, but the RAPS mitochondria were prepared by a method which avoids the addition of artefacts during the dehydration and embedding of the tissue. Although the fine structures of the mitochondria are not seen by this latter method, the major mitochondrial features are present in both electron micrographs. This shows that these are not preparation artefacts. Figure 1.3(*b*) is a diagrammatic representa-

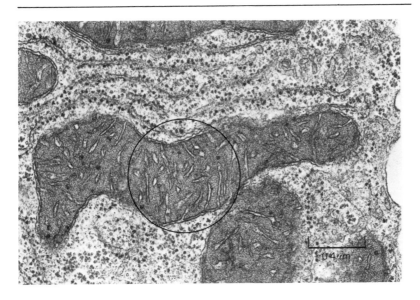

Fig. 1.3(a)   Electron micrograph of rat liver mitochondrion. Glutaraldehyde and osmium tetroxide fixed, stained with uranyl acetate and lead citrate. (Courtesy of Dr G. Bullock.)

tion of part of the rat liver mitochondrion shown in Fig. 1.3(a). The outer membrane is usually rather smooth and featureless (Fig. 1.3(d)), although very high resolution electron micrographs suggest the presence of pits and pores approximately 25–30 Å in diameter. In contrast the inner membrane, which for the most part tends to follow the contour of the outer membrane, in places invaginates to produce structures known as cristae mitochondriales (more usually – cristae). Mitochondria from different sources exhibit a wide variety in number and shape of cristae. Typically, mitochondria from highly active tissues, e.g. insect or bird flight muscle and heart muscle, have large numbers of tightly packed cristae (Fig. 1.4(a) and (b)). At the other end of the scale mitochondria of many parasites of the alimentary tract, such as liver flukes or tapeworms, which, because of their almost anaerobic environment can be capable of only limited respiratory activity, contain relatively few cristae (Fig. 1.5).

Although in most animal mitochondria the cristae tend to be clearly defined laminar structures, in plants (Fig. 1.6(a)) and some animal mitochondria, e.g. rat adrenal cortex (Fig. 1.6(b)), they are usually more tenuous finger-like processes. The cristae are usually arranged transverse to the long axis of the mitochondria, sometimes extending across the short

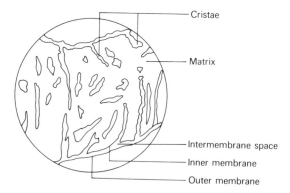

Fig. 1.3(b) Diagrammatic representation of structure of the ringed part of the rat liver mitochondrion shown in Fig. 1.3a.

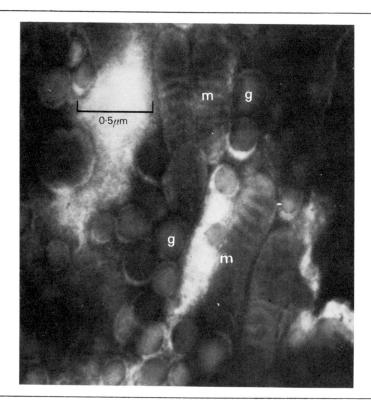

Fig. 1.3(*c*)   Frozen section of a part of a rat anterior pituitary somatotroph. The tissue was fixed in glutaraldehyde before freezing in liquid nitrogen and sectioning at —80°C. This procedure avoids the possibility of artefacts introduced during dehydration and embedding of the tissue. The structure of the mitochondria (m) and storage granules (g) is clearly seen in material prepared in this way. Stained with 4 per cent silicotungstic acid. (Courtesy of Dr S. Howell.)

Fig. 1.3(*d*)   Electron micrograph of freeze-etch preparation of isolated rat liver mitochondrion. The mitochondria were unfixed and were etched for 2 min. at —100°C. The encircled arrow indicates the direction of shadowing. The three arrows point out a faint circular patch. (Courtesy of Dr J. M. Wrigglesworth.)

axis of the mitochondria, sometimes only part of the way across. Insect flight muscle mitochondria contain cristae with septa (Fig. 1.7). The regularly arranged perforations are called fenestrations. In negatively stained preparations of mitochondrial inner membrane (in which a thin film of an electron-opaque salt such as phosphotungstate is dried around the specimen leaving the unpenetrated part of the specimen electron-transparent) the inward facing surface is seen to contain an array of regularly spaced stacked particles (Fig. 1.8 (*a*)–(*c*)). These particles, sometimes called Fernandez-Moran particles (after their discoverer) or tripartite repeating units, seem to comprise three sections — a more or less spherical head piece, a cylindrical stalk and a base piece (sect. 3.8). The dimensions of these particles have been measured by various workers in different

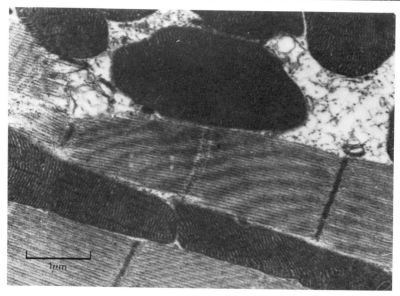

Fig. 1.4(*a*)   Mitochondria from *Musca domestica* (housefly) flight muscle showing tightly packed cristae. Glutaraldehyde/osmium tetroxide fixed; uranyl acetate/lead citrate stained. (Courtesy of Dr M. Tribe and Mrs S. Webb.)

mitochondria; those shown in Fig. 3.29 are typical of stalked particles from a number of species.

Although these particles have been demonstrated using several different negative staining techniques, they are not usually seen in sections of intact mitochondria and it has been suggested that they may be artefacts resulting from the negative staining procedure. Even if this were true, which is by no means certain, the regular shape and distribution of the membrane particles is a clear indicator of a macromolecular repeating structure located close to the matrix facing surface of the inner membrane.

The two membranes of the mitochondria delimit two mitochondrial spaces (see Fig. 1.3(*b*)). The intermembrane and intracristal spaces are almost certainly continuous although in electron micrographs of some types of mitochondria cristae may appear to be vesicular, which would indicate a discontinuity. However, careful serial sectioning through mitochondria shows that such cristae are attached to the inner membrane by small tubes (pediculi) which may not appear in individual sections (Fig. 1.9). It is possible, however, that the intracristal and intermembrane spaces could be made temporarily distinct by closure of the pediculi. The intermembrane and intracristal spaces appear to be relatively unstructured with little or no granular inclusions. The inner membrane and its invaginations surround the matrix space, which usually has a granular appearance.

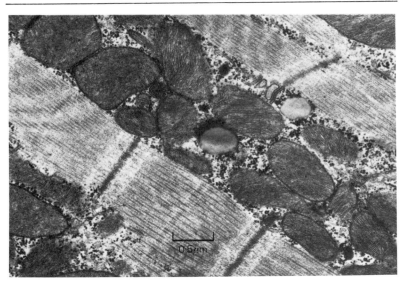

Fig. 1.4(*b*)   Electron micrograph of rat cardiac muscle mitochondria.
Glutaraldehyde and osmium tetroxide fixed, stained with uranyl acetate and lead
citrate. (Courtesy of Dr G. Bullock.)

Although the nature of the granules is not clearly established it is possible
that they are sites of calcium deposition. There is some evidence that the
release of large amounts of calcium from the mitochondria may be an
important regulatory mechanism for cellular metabolism (sect. 4.9.2).
Other important inclusions of the mitochondrial matrix are ribosomes,
which we shall see (sect. 5.3) are usually distinguishable from the extramitochondrial ribosomes and are found closely associated with the matrix
facing surface of the inner membrane and cristae. The matrix also contains
DNA which is usually attached to the inner membrane at a small number
of points.

As emphasised at the end of the last section the mitochondrion should
not be viewed as a static unchanging object. The appearance of mitochondria can vary considerably during development and differentiation of cells,
in response to differing dietary and growth conditions and as a result of
pathological disorders. The appearance of isolated mitochondria is very
much dependent on the suspension medium and the metabolic state of the
organelle.

1.3   Functional organisation of mitochondria

1.3.1 Preparation of mitochondria
The main approach to studying the functional organisation of mitochondria has been to separate the component parts of the mitochondria and to

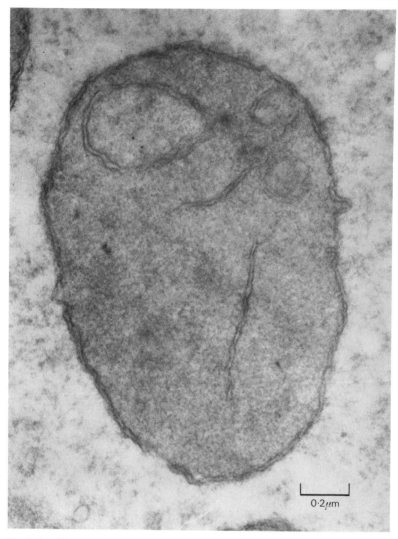

Fig. 1.5 Electron micrograph of mitochondrion from liver fluke parenchymal cell. The preparation was fixed in osmium tetroxide and stained with uranyl acetate. (Courtesy of Dr N. Björkman.)

examine the composition and properties of the different fractions. It is initially necessary to isolate the mitochondria in as unchanged a state as possible and free of contaminating extramitochondrial material. The tech-

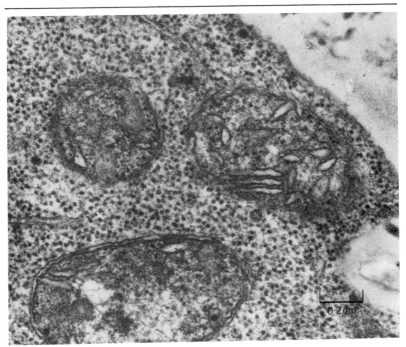

Fig. 1.6(a)   Mitochondria from root-tip of *Suaeda maritima*. Glutaraldehyde and osmium fixed. (Courtesy of Dr J. Hall. From J. Hall, T. J. Flowers and R. M. Roberts, *Plant Cell Structure and Metabolism*, Longman, 1974.)

nique which is usually used is based on that developed in 1948 by Schneider and Hogeboom using rat liver mitochondria. The first step after the preliminary dicing of the tissue to suitably sized pieces is to break the cells. This is carried out at about $0°-4°C$ in an isotonic suspension medium — sucrose or mannitol are usually used. Liver and kidney can be broken by gentle homogenisation using a glass mortar with a Teflon pestle with a 0.1 mm clearance. In the case of tougher animal tissues such as heart muscle it is also necessary to soften the tissue by treatment with a bacterial proteolytic enzyme (nagarse) before satisfactory cell breakage can be achieved. Similarly with plants and yeast it may be necessary to digest the cell wall with a carbohydrase before gentle breakage of the cell membrane is possible. If enzymatic treatment has been used it is necessary at this stage to subject the homogenisation to centrifugation at 8 000 – 10 000 g for 10 min. The supernatant, containing ribosomes, endoplasmic reticulum and cytosol as well as the added enzyme, is discarded and the sediment is suspended in fresh isolation medium lacking the enzyme. After

11

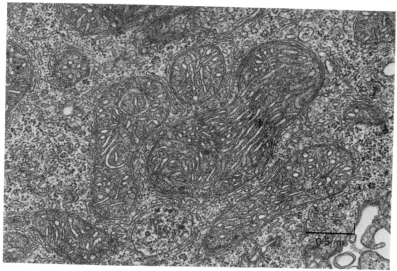

Fig. 1.6(*b***)**   Electron micrograph showing rat adrenal cortex mitochondria. Glutaraldehyde and osmium tetroxide fixed, stained with uranyl acetate and lead citrate. (Courtesy of Dr G. Bullock.)

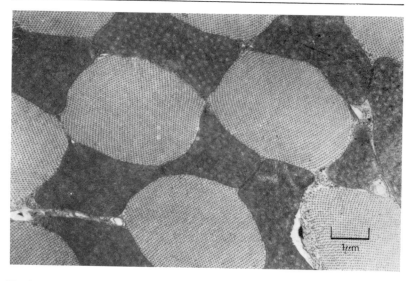

Fig. 1.7   Transverse section of mitochondria from *Calliphora erythrocephala* (blowfly) flight muscle. Glutaraldehyde/osmium tetroxide fixed; uranyl acetate/lead citrate stained. (Courtesy Dr M. Tribe and Mrs S. Webb.)

enzyme treatment this suspension in common with untreated homogenates is given a relatively slow spin (600 g for 10 min.) to sediment unbroken cells, cell debris and nuclei, and the supernatant is spun at 8 000–10 000 g for 10 min. to give a pellet containing mitochondria. A further one or two 8 000–10 000 g spins are usually employed to wash the mitochondria free of contaminating non-mitochondrial material. These preparations are summarised in the flow diagram of Fig. 1.10. Mitochondria isolated in this way have been used for most of the studies on oxidative phosphorylation, membrane permeability, protein synthesis and other aspects of mitochondrial biochemistry.

1.3.2 Localisation of mitochondrial components

To establish firmly the locations of particular mitochondrial activities it is necessary to fractionate mitochondria further. In recent years techniques have been developed for removing the outer membrane. These involve either swelling of the mitochondria in hypotonic buffer or treatment with low concentrations of the detergent digitonin, which selectively detach the outer membrane – the separation can often be improved by mild sonication. The different fractions of the mitochondria can be isolated by centrifugation as a sucrose gradient. A very heavy fraction (density c. 1.20 g cm^{-3}) can be identified as the inner membrane by the presence of tripartite membrane particles in electron micrographs after negative staining. The light fraction (density c. 1.10 g cm^{-3}) which lacks the membrane particles is presumed to be the outer membrane. During the initial removal of the outer membrane a soluble fraction is released which includes the contents of the intermembrane space and part of the matrix released from damaged inner membrane vesicles.

Examination of the enzyme activities of the two membrane fractions shows that the inner membrane is the site of electron carriers (including succinate dehydrogenase) and ATP synthesis. As we shall see later (sect. 3.8) the ATP synthesising complex is almost certainly located in the head pieces of the inner membrane particles. Other important enzymes which have been located on the inner membranes are an electron-transport-linked NAD:NADP transhydrogenase (EC.1.6.1.1., sect. 3.10) and some of the enzymes involved in fatty acid elongation (for C_{10} fatty acids). The enzymes found in the outer membrane fraction are a functionally heterogeneous group and it has not usually been possible to provide a rationale for their location on this membrane. They include monamine oxidase (EC.1.4.3.4.), which is involved in the oxidation of adrenaline, serotonin and tryptamine; kynurenine hydroxylase (EC.1.14.13.9.) – involved in typtophan degradation; and a number of enzymes involved in phospholipid metabolism. A fatty acyl CoA synthetase (EC.6.2.1.3.) and the system for elongation of C_{14} and C_{16} fatty acids are also located in the outer membrane.

The presence of an enzyme in the soluble fraction indicates that it

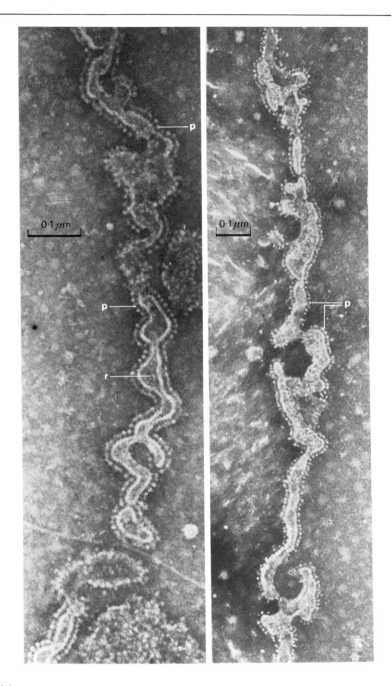

Fig. 1.8   Tripartite membrane particles from mitochondrion of flight muscle of *Calliphora erythrocephala*. (*a*) Mitochondria sonicated 15 sec. stained with 2 per cent sodium phosphotungstate. (*b*) Osmotically lysed mitochondria stained with 2 per cent sodium silicotungstate. Each contains a ribbon (r) which is believed to represent a tubular breakdown product of cristae. These are studded with the particles (p). (Courtesy of Dr E. A. Munn.)

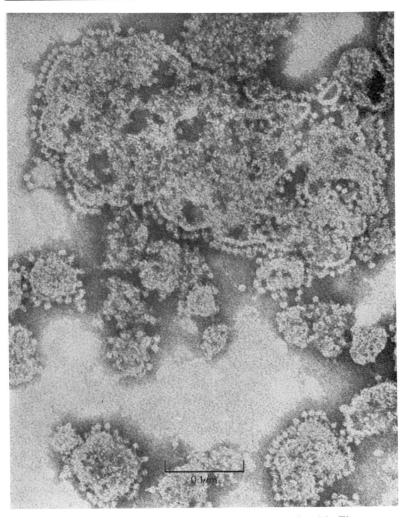

Fig. 1.8(*c*)   Submitochondrial particles from beef heart mitochondria. The preparation was negatively stained with phosphotungstate. Tripartite membrane particles are clearly visible. (Courtesy by Prof E. Racker.)

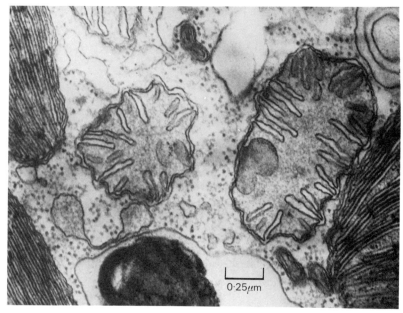

Fig. 1.9 Mitochondria from *Euglena spirogyra*. Fixed in osmium tetroxide, stained by lead hydroxide. The intracristal space and intermembrane space are joined by pediculi. (Courtesy of Biophoto Associates/Dr G. F. Leedale.)

originates from either the intermembrane space or the matrix. Matrix enzymes can be identified as they also appear in the heavy fraction (trapped within any inner membrane vesicles). Enzymes found in the matrix include those of the tricarboxylic acid cycle, part of the urea cycle (in ureotelic animals) and of fatty acid oxidation. In addition, the enzyme systems for synthesis of mitochondrial DNA and RNA, and those proteins manufactured within mitochondria are present in the matrix. The enzyme superoxide dismutase (EC.1.15.1.1.) responsible for rapid removal of the radical intermediates of oxygen reduction has recently been shown to be present in the matrix. The enzymes nucleoside diphosphokinase (EC.2.7.4.6.) which catalyses the transfer of the terminal phosphate to a nucleoside diphosphate, and adenylate kinase (EC.2.7.4.3.) which transfers the terminal phosphate of ATP to AMP, appear only in the soluble fraction and are consequently located in the intermembrane space.

The membranes of mitochondria differ considerably in their permeability properties. The outer membrane appears to be more or less freely permeable to small or medium sized molecules or ions. On the other hand the inner membrane, possibly because of its higher cardiolipin content, is permeable only to small uncharged molecules such as O_2 or undissociated

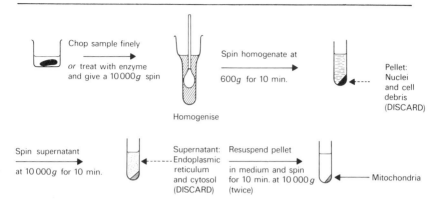

Fig. 1.10   Flow diagram showing the preparation of mitochondria. All the procedures are carried out at 4°C and using isotonic cold buffer at ~pH 7.4.

H_2O. Nevertheless mitochondrial function does require that certain metabolites do cross the mitochondrial membrane and as we shall see in Chapter 4 this membrane contains a number of carrier enzymes or enzyme systems whose function is to transport metabolites through the membrane. The approach to identification of these carriers and demonstration of their location involves study of the penetration of the mitochondria by metabolites using labelling studies or mitochondrial swelling, use of specific inhibitors of certain carriers and comparison of the effects of addition of metabolites on some mitochondrial parameter (such as oxygen uptake) to intact and fragmented mitochondria.

1.4   ATP and biological work

Although mitochondria exhibit a variety of different functions there is no doubt that the most important one is electron transport linked to ATP synthesis (oxidative phosphorylation), the structure of ATP is shown in Fig. 1.11. It is likely that most other mitochondrial functions, transport phenomena, DNA, RNA, protein synthesis, tricarboxylic acid cycle and β-oxidation, either in whole or in part play a supportive role to the ATP synthesising system. Before embarking on a full discussion of mitochondrial function it is of value to consider the involvement of ATP in biological processes.

The work carried out by living organisms is generally of three kinds:

(a) *Mechanical work* — muscle contraction, movement of flagellae and cilia, contraction of microtubules etc.;

(b) *Osmotic work* — establishment and maintenance of concentration gradients both between the inside and the outside of the cell and between different intracellular compartments;

17

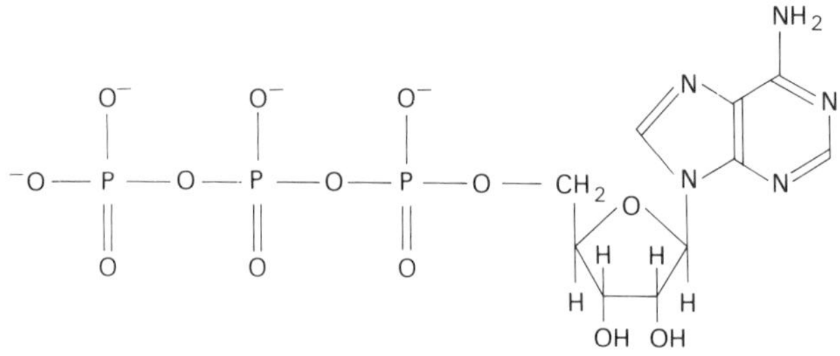

Fig. 1.11   The structure of adenosine 5' triphosphate (ATP).

(c) *Biosynthetic work* – elaboration of the complex molecules necessary to the structure and function of the cell.

Considering muscle contraction as an example of mechanical work, the sliding of the actin and myosin filaments of the myofibril is accompanied by the hydrolysis of ATP to ADP and inorganic phosphate (ATPase activity). The precise sequence of biochemical events is not yet fully established, but it is thought that ATP initially reacts with the globular head piece of myosin in the actomyosin complex. This is followed by breakage of the actin myosin link and the hydrolysis of ATP. This is thought to result in a conformational change in the myosin molecule such that it can re-form a link with the actin at another point, the ADP and phosphate leave the myosin. The overall results of the process are contraction of actomyosin and ATP hydrolysis. A very similar situation exists in the so-called sodium pump in the cell membrane, which has been extensively studied in nerve axon and red blood cells. The pump, which is involved in maintaining the high intracellular concentration of K^+ and low intracellular concentration of Na^+ (compared with the surrounding fluid), operates by pumping K^+ into the cell in exchange for the movement of Na^+ outwards. This exchange, operating against two concentration gradients, is accompanied by ATP hydrolysis. The ATP transfers its terminal phosphate residue to the membrane carrier and ultimately the enzyme donates this phosphate to water. In the nerve axon the sodium pump is important in re-establishing the K^+ and Na^+ gradients which are diminished by the inward movement of Na^+ during the passage of an action potential and the compensatory outward movement of K^+. The sodium and potassium gradients established by the sodium pump have another important function, they are used to drive the uptake of glucose, amino acids and possibly other metabolites across the cell membrane against concentration gradients, using different specific carriers. ATP is probably not directly involved here, but the ATP-driven sodium pump is clearly essential to continued active transport of these metabolites.

ATP is implicated in a large number of biosynthetic reactions. Some of these involve an overall hydrolysis of ATP and ADP and inorganic phosphate as in muscle contraction and the sodium pump. An example is the synthesis of glutamine from glutamate and ammonia, catalysed by glutamine synthetase (EC.6.3.1.2.). The reaction is:

Glutamate + NH_3 + ATP $\rightleftharpoons$ Glutamine + ADP + Pi

It is thought that the ATP donates its terminal phosphate to glutamate giving glutamyl phosphate which then reacts with ammonia giving glutamine and releasing inorganic phosphate. ATP is similarly involved in carbamoyl phosphate synthesis from CO_2 and NH_3 (important in the biosynthesis of pyrimidines and urea) and also in a number of steps in the biosynthesis of purines. GTP (guanosine triphosphate) plays a similar role in addition of amino acids to a growing polypeptide chain.

The overall hydrolysis of ATP to ADP and inorganic phosphate accompanying many biological work processes has led to the somewhat misleading statement that ATP hydrolysis drives biological work processes. It is important to note that ATP hydrolysis to ADP and Pi does not occur as a single reaction in any of the preceding situations. The primary reaction is usually the donation of phosphate by ATP to an acceptor — either an enzyme or a substrate. The phosphorylated intermediate then transfers the phosphate to water in a reaction obligatorily coupled under physiological conditions to the work process. This means that tables of standard free energies of hydrolysis of phosphates are only really useful in so far as they provide a measure of the phosphate donor potential to the standard acceptor (water). Table 1.1 shows that the phosphate donor potential of ATP is intermediate between those compounds usually used in ATP synthetic reactions, e.g. phosphoenolpyruvate and 1,3-diphosphoglycerate, which have high phosphate donor potential, and those whose synthesis involves phosphate transfer from ATP, e.g. glucose-6-phosphate, which has low phosphate donor potential.

Phosphate transfer is not the only type of reaction in which ATP participates, however, it can act as a pyrophosphate donor as in the synthesis of phosphoribosyl pyrophosphate:

ATP + D-ribose-5-phosphate $\rightleftharpoons$ AMP + phosphoribosyl pyrophosphate

Table 1.1   Standard free energy of hydrolysis of some phosphorylated compounds

Phosphorylated compound	$\Delta G^{O'}$	
	$kJ mol^{-1}$	($kcal\ mol^{-1}$)
Phosphoenolpyruvate	−61.9	(−14.8)
1,3-Diphosphoglycerate	−49.4	(−11.8)
ATP	−30.5	(− 7.3)
Glucose-6-phosphate	−13.8	(− 3.3)

or as an AMP donor as in the activation of an amino acid prior to transfering the amino acid to a specific transfer RNA molecule. The reaction is catalysed by an aminoacyl tRNA synthetase, relevant to the particular amino acid:

$RCHNH_2COOH + ATP \rightleftharpoons RCHNH_2CO\text{-}AMP + PPi$ (pyrophosphate)
$RCHNH_2CO\text{-}AMP + tRNA \rightleftharpoons RCHNH_2CO\text{-}tRNA + AMP$

A number of processes similarly involve ATP or other nucleoside triphosphates, e.g. fatty acid activation (ATP), phospholipid synthesis (CTP), glycogen synthesis (UTP) and starch synthesis (ATP). A special case of this type of reaction is the transfer of nucleoside monophosphate from nucleoside triphosphate in nucleic acid synthesis.

Clearly there is an enormous cellular demand for nucleoside triphosphate. Under aerobic conditions by far the greatest proportion of this demand is satisfied by ATP synthesis from ADP and inorganic phosphate in mitochondria, an illustration of the importance of this organelle. The mitochondrial oxidative phosphorylation system is very specific as it will not bring about phosphorylation of significant quantities of other nucleoside diphosphates other than ADP, or the phosphorylation of AMP. In this respect the two enzymes we have seen to be present in the intermembrane space play important roles.

Nucleoside diphosphokinase (EC.2.7.4.6.) catalyses the transfer of the terminal phosphate of ATP to a nucleoside diphosphate, for example:

$ATP + GDP \rightleftharpoons ADP + GTP$
or $ATP + deoxyADP \rightleftharpoons ADP + deoxyATP$

Adenylate kinase (EC.2.7.4.3.) catalyses the transfer of the terminal phosphate of ATP to AMP, the reaction is as follows:

$ATP + AMP \rightleftharpoons 2ADP$

In either case the ADP formed can be taken up into the mitochondrial matrix and converted into ATP by oxidative phosphorylation (Ch. 3).

Suggested further reading

Books
LEHNINGER, A. L. (1965) *The Mitochondrion. Molecular Basis of Structure and Function.* Benjamin, New York and Amsterdam.
LEHNINGER, A. L. (1971) *Bioenergetics* (2nd edn.). Benjamin, Menlo Park, California.
MUNN, E. A. (1974) *The Structure of Mitochondria.* Academic Press, London and New York.
TRIBE, M. A. and WHITTAKER, P. A. (1972) *Chloroplasts and Mitochondria. Studies in Biology,* No. 31. Edward Arnold, London.

Methods

CHAPPELL, J. B. and HANSFORD, R. G. (1972) Preparation of mitochondria from animal tissues and yeast, in *Subcellular Components; Preparation and Fractionation* (2nd edn.), ed. G. D. Birnie. Butterworth, London.

GRIMSTONE, A. V. (1968) *The Electron Microscope in Biology. Studies in Biology* No. 9. Edward Arnold, London.

HOGEBOOM, G. H., SCHNEIDER, W. C. and PALLADE, G. E. (1947) The isolation of morphologically intact mitochondria from rat liver, *Proc. Soc. Exp. Biol. and Med.*, **65**, 320–1.

Other topics

DAHL, J. L. and HOKIN, L. E. (1974) The sodium-potassium ATPase, *Ann. Rev. Biochem.*, **43**, 327–56.

HOFFMAN, H. P. and AVERS, C. J. (1973) Mitochondrion of yeast: Ultrastructural evidence for one giant branched organelle per cell, *Science*, **181**, 749–51.

KENNEDY, E. P. and LEHNINGER, A. L. (1949) Oxidation of fatty acids and tricarboxylic acid cycle intermediates by isolated rat liver mitochondria, *J. Biol. Chem.*, **179**, 957–72.

MANTON, I. (1959) Electron microscopic observations on a very small flagellate: The problem of *Chromulina pusilla* Butcher, *J. Mar. Biol. Ass. U.K.*, **38**, 319–33.

SCHNAITMAN, C. and GREENAWALT, J. W. (1968) Enzymatic properties of the inner and outer membrane of rat liver mitochondria, *J. Cell Biol.*, **38**, 158–75.

WRIGGLESWORTH, J. M., PACKER, L. and BRANTON, D. (1970) Organisation of mitochondrial structure as revealed by freeze etching, *Biochim. Biophys. Acta*, **205**, 125–35.

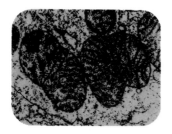

Chapter 2

Mitochondrial metabolism

2.1 Introduction

In Chapter 1 it was suggested that the phosphorylation of ADP to ATP associated with the inner membrane was the primary function of mitochondria. This chapter considers some of the metabolic processes of mitochondria, especially those which play a crucial role in ATP synthesis. On the whole these processes occur in the matrix of the mitochondrion, the enzymes responsible being within the space surrounded by the inner membrane. Here they generate reducing potential, which is the driving force for the electron transport system and ADP phosphorylation within this membrane. The operation of electron transport is initiated by the reduction, by a metabolite from the matrix, of either nicotinamide adenine dinucleotide (NAD^+) or flavin adenine dinucleotide (FAD), which are associated with an oxidoreductase enzyme (see sect. 3.2). Two main metabolic processes generate the reductive metabolites, these are the tricarboxylic acid cycle (abbreviated to the TCA cycle and also referred to as the Krebs or citric acid cycle) and the β-oxidation pathway for fatty acids. However, it would be incorrect to think that the sole purpose of these metabolic processes is to generate ATP. The TCA cycle, for example, provides biosynthetic precursors for a number of important biomolecules. The use of these precursors leads to a depletion of intermediates of the cycle which is made good by a number of anaplerotic or 'topping-up' reactions.

Full accounts of the enzymes of the TCA cycle and β-oxidation, including the history of their discovery and experimental evidence for their existence, can be found in all general biochemistry texts and will not be given here. The aim of this chapter is to emphasise the overall action of the metabolic processes associated with the mitochondria, their organisation, interaction with other metabolic pathways and their integration and control.

2.2 The tricarboxylic acid cycle

The reactions of the TCA cycle are shown in Fig. 2.1. The cycle is initiated by the condensation of acetyl CoA and oxaloacetate to give citrate (reaction 2) and the citrate is metabolised by a number of oxidation and decarboxylation reactions to regenerate oxaloacetate. Effectively the acetate originally donated by acetyl CoA to the cycle is completely oxidised to CO_2 and H_2O by a single turn of the cycle. In fact the two carbon atoms which appear as CO_2 after a single turn of the cycle are not the two carbon atoms donated by the acetyl CoA. This has been shown by adding radiolabelled acetate and monitoring the label in the CO_2 given off. This is illustrated in Fig. 2.1, the two carbons of the acetate are in dark type and after one turn of the cycle these radiolabelled carbons are in the oxaloacetate and not the CO_2. However, the overall or *net* effect is the complete oxidation of the acetyl group. The carbon dioxide is released at reactions 4 and 5, those catalysed by NAD^+-linked isocitrate dehydrogenase and the 2-oxoglutarate* dehydrogenase complex (sect. 2.5).

In the operation of the cycle, starting from pyruvate, five oxidation reactions occur (reactions 1, 4, 5, 7 and 9 in Fig. 2.1). These oxidation reactions are coupled to the reduction of NAD^+ to NADH in the reactions catalysed by pyruvate, isocitrate, 2-oxoglutarate and malate dehydrogenases and to the reduction of FAD to $FADH_2$ in the case of succinate dehydrogenase. The reoxidation of each molecule of NADH and $FADH_2$ by oxygen via the electron transport system (Ch. 3) leads to the phosphorylation of three and two molecules of ADP respectively. The reaction catalysed by succinic thiokinase (reaction 6, Fig. 2.1) can give rise to another molecule of ATP. The phosphorlysis of the thioester bond of succinyl CoA is followed by the phosphorylation of GDP to GTP, and GTP can be used to phosphorylate ADP, a reaction catalysed by nucleoside diphosphokinase (sect. 1.4). Therefore the oxidation of acetyl CoA to CO_2 and H_2O by one turn of the cycle can give rise to 12 ATPs and the oxidation of pyruvate to CO_2 and H_2O can give rise to 15 ATPs.

Complete oxidation of pyruvate to CO_2 and H_2O occurs with a standard free energy change (at pH 7.0 and 25°C) of $-1\,422\,kJ\,mol^{-1}$ (340 kcal mol^{-1}). The standard free energy change (at pH 7.0 and 25°C) accompanying phosphorylation of ADP to ATP is $+30.5\,kJ\,mol^{-1}$

---

*2-Oxoglutarate will be used throughout this book for the metabolite also called α-ketoglutarate.

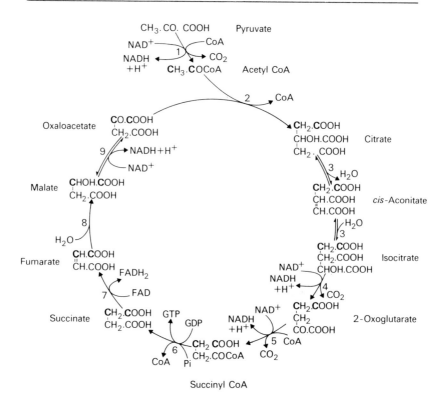

Fig. 2.1   THE TRICARBOXYLIC ACID CYCLE.
1. Pyruvic dehydrogenase complex. 2. Citric synthase (EC.4.1.3.7.). 3. Aconitase
(EC.4.2.1.3.). 4. Isocitrate dehydrogenase (EC.1.1.1.41.). 5. 2-Oxoglutarate
dehydrogenase complex. 6. Succinic thiokinase (EC.6.2.1.4.). 7. Succinate
dehydrogenase (EC.1.3.99.1.). 8. Fumarase (EC.4.2.1.2.). 9. Malate dehydrogenase
(EC.1.1.1.37.). C represents radio(^{14}C)-labelled carbon derived from added acetate.

(7.3 kcal mol^{-1}). Consequently the synthesis of 15 moles of ATP during
pyruvate   oxidation   requires   a   total   free   energy   input   of
15 × 30.5 kJ = 457.5 kJ. This represents an efficiency of

$$\frac{457.5}{1\,422} \times 100 = 32 \text{ per cent}$$

which compares favourably with many non-biological processes. It should
be remembered that these figures are calculated assuming molar concentrations of all reactants — a situation which is never achieved in the cell.

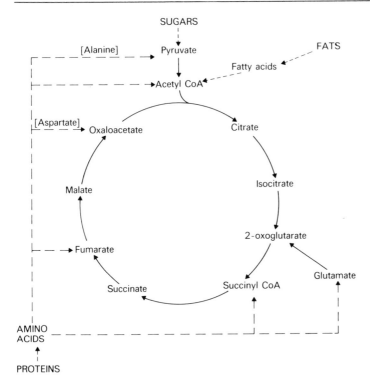

Fig. 2.2 Degradative pathways leading into the TCA cycle.

2.3   The central role of the TCA cycle

Figure 2.2 briefly summarises the degradative pathways which lead into
the TCA cycle. Clearly the TCA cycle plays a central role in metabolism,
as sugars, fats and amino acids are all degraded to intermediates of the
TCA cycle. Sugar catabolism (glycolysis) occurs in the cytosol; its end-
product is pyruvate which is transported into the mitochondria (sect. 4.4)
where it undergoes oxidative decarboxylation to give acetyl CoA, which
can then enter the TCA cycle. The degradative pathway for fatty acids is
β-oxidation and the enzymes for this pathway are located entirely within
the mitochondrial matrix. The product from β-oxidation of most fatty
acids is acetyl CoA. Amino acid degradation does not occur via a unified
pathway and the end-products of their preliminary breakdown can be
pyruvate, acetyl CoA, oxoglutarate, succinyl CoA, fumarate or oxaloacetate. As a general rule the rather complex pathways for the degradation

25

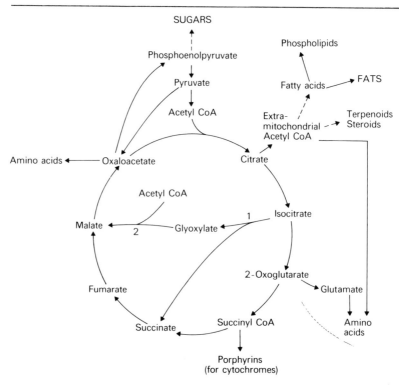

Fig. 2.3 The TCA cycle as a source of precursors for biosynthesis.
1. Isocitrate lyase. 2. Malate synthase.

of most amino acids occur outside the mitochondria and one of the final products in each pathway is transported into the mitochondria to be oxidised by the TCA cycle.

The interaction of the TCA cycle with some biosynthetic pathways is briefly summarised in Fig. 2.3. The cycle provides precursors for the synthesis of sugars, fats and amino acids. Sugar biosynthesis (gluconeogenesis) occurs partly by the reversal of the glycolytic sequence. However, some of the reactions of glycolysis are not readily reversible in the direction of glucose synthesis and are bypassed by alternative processes. One of the irreversible reactions is the conversion of phosphoenolpyruvate to pyruvate by the enzyme pyruvate kinase (EC.2.7.1.40) in the cytosol of liver. Figure 2.4 shows the reaction sequence involving enzymes in the matrix and the cytosol, which converts pyruvate into phosphoenolpyruvate. The matrix enzymes pyruvate carboxylase (EC.6.4.1.1) and malate dehydrogenase convert pyruvate into malate. The malate is then transported out of

CYTOSOL   MATRIX

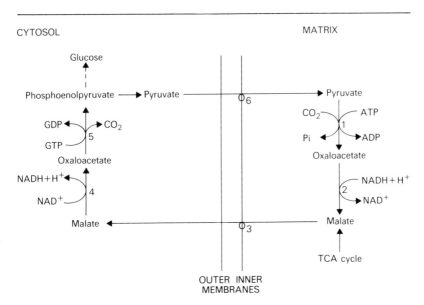

Fig. 2.4   The formation of phosphoenolpyruvate for gluconeogenesis.
1. Pyruvate carboxylase. 2. Mitochondrial malate dehydrogenase. 3. Dicarboxylate transporter. 4. Cytosolic malate dehydrogenase. 5. Phosphoenolpyruvate carboxykinase. 6. Pyruvate transporter.

the matrix by the dicarboxylate transporting system (sect. 4.4) and is then converted by extramitochondrial malate dehydrogenase and phosphoenolpyruvate carboxykinase (EC.4.1.1.32) into phosphoenolpyruvate (PEP). The PEP can then be used as a precursor for glucose biosynthesis in the cytosol. The overall mechanism is the conversion of pyruvate to oxaloacetate to phosphoenolpyruvate. In rat and mouse liver, PEP carboxykinase is in the cytosol and so the mechanism shown in Fig. 2.4 will operate, but in some tissues, e.g. rabbit and pigeon liver, PEP carboxykinase is in the mitochondrial matrix and a different mechanism may operate. The pyruvate would then be converted to oxaloacetate and then PEP in the matrix, the PEP may be transported out of the matrix on the tricarboxylate transporter (sect. 4.4).

The biosynthesis of fatty acids takes place outside the mitochondria and is catalysed by the enzyme complex, fatty acid synthetase; the details of this will not be considered here. The precursor for the synthesis of fatty acids, terpenoids and steroids is acetyl CoA. Acetyl CoA cannot cross the mitochondrial inner membrane, therefore the transfer of the intramitochondrial acetyl group into the cytosol is achieved by the reaction sequence shown in Fig. 2.5. Acetyl CoA condenses with oxaloacetate to give citrate, which is transported out of the matrix on the tricarboxylate

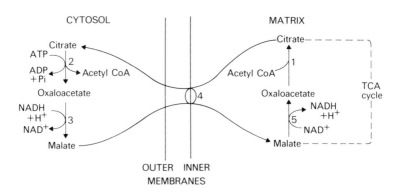

Fig. 2.5   The transfer of acetyl groups out of the matrix.
1. Citrate synthase. 2. Citrate lyase. 3 and 5. Malate dehydrogenase. 4. Tricarboxylate transporter.

transporter (Ch. 4). Acetyl CoA is regenerated in the cytosol by the enzyme ATP-citrate lyase (EC.4.1.3.8). The oxaloacetate can be returned to the matrix by its conversion to malate, which exchanges with outgoing citrate on the tricarboxylate transporter. Once in the matrix malate can be reoxidised to give oxaloacetate.

As with amino acid degradative pathways, biosynthesis of each amino acid occurs by an individual metabolic pathway. Many eukaryotes, of course, are incapable of making all amino acids (those which are not made are termed essential, and must be included in the diet). Wherever synthesis does occur either phosphoenolpyruvate, pyruvate or a TCA cycle intermediate contributes to the carbon skeleton of the amino acid. Amino acid biosynthesis also occurs outside the mitochondria and the precursors have to be transported into the cytosol.

Some important enzymes which connect amino acid metabolism with the TCA cycle are glutamate dehydrogenase (EC.1.4.1.2), aspartate amino transferase (EC.2.6.1.1) and alanine amino transferase (EC.2.6.1.2). All three enzymes are involved in amino acid synthesis and breakdown. Glutamate dehydrogenase, a mitochondrial matrix enzyme catalyses the oxidative deamination of glutamate to yield oxoglutarate and ammonia, the reaction is:

$$\text{Glutamate} + \text{NAD}^+ \rightleftharpoons \text{Oxoglutarate} + \text{NADH} + \text{NH}_4^+$$

This reaction converts the carbon skeleton of glutamate into oxoglutarate which can be oxidised by the TCA cycle. The ammonia can be used for urea synthesis (sect. 2.6). The reverse reaction is the conversion of NH_3

into the amino group of glutamate. This reaction can be toxic to the cell as in the presence of excess ammonia, oxoglutarate will be converted to glutamate thereby preventing the operation of the TCA cycle. This could be disastrous for the brain, which synthesises ATP almost exclusively using glycolysis and the TCA cycle.

The amino group of glutamate can be transferred to other amino acids (or vice versa) by the action of an aminotransferase. These are pyridoxal phosphate containing enxymes which catalyse the following reversible reaction:

2-Oxo acid + Glutamate $\rightleftharpoons$ Amino acid + 2-Oxoglutarate

Two aminotransferases were mentioned earlier, aspartate amino transferase and alanine amino transferase. Both enzymes occur in the cytostol and matrix, and they catalyse the following reactions:

Aspartate + Oxoglutarate $\rightleftharpoons$ Glutamate + Oxaloacetate
Alanine + Oxoglutarate $\rightleftharpoons$ Glutamate + Pyruvate

These two enzymes are sometimes called glutamic oxaloacetic transaminase (GOT) and glutamic pyruvic transaminase (GPT) respectively.

The action of glutamate dehydrogenase and the aminotransferases allows the complete oxidation of the carbon skeleton of amino acids via the TCA cycle, or converts some TCA cycle intermediate into the carbon skeleton of an amino acid.

2.4   Anaplerotic reactions

Removal of TCA cycle intermediates for sugar, lipid or amino acid synthesis means that not all of the oxaloacetate which condenses with acetyl CoA is replenished by the cycle. If no provision was made for replacing the oxaloacetate from another source the cycle would ultimately stop. There are a number of reactions (anaplerotic reactions) which carry out this 'topping-up' function. An important enzyme for replenishing oxaloacetate is the matrix enzyme pyruvate carboxylase which, as shown in Fig. 2.4, is also involved in gluconeogenesis. This biotin-containing enzyme is capable of forming oxaloacetate by the carboxylation of pyruvate in the presence of ATP. The reaction is:

Enzyme-biotin + ATP + CO_2 + H_2O $\rightleftharpoons$ Enzyme-carboxybiotin + ADP + Pi
Enzyme-carboxybiotin + Pyruvate $\rightleftharpoons$ Enzyme-biotin + Oxaloacetate

Overall:

Pyruvate + ATP + CO_2 + H_2O $\rightleftharpoons$ Oxaloacetate + ADP + Pi

This enzyme has an absolute requirement for acetyl CoA, if the acetyl CoA level is high pyruvate carboxylase will be stimulated.

29

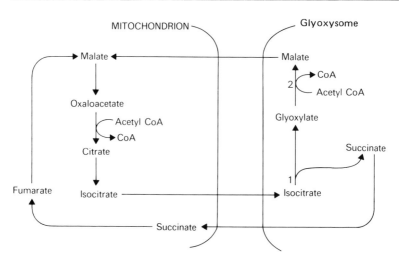

Fig. 2.6   The glyoxylate bypass.
1. Isocitrate lyase (EC.4.1.3.1.). 2. Malate synthase (EC.4.1.3.2.).

In higher plants and microorganisms, but not animals, the glyoxylate bypass (shown in Fig. 2.3) may play an anaplerotic role. The two enzymes of the bypass, isocitrate lyase and malate synthase (reactions 1 and 2 in Fig. 2.6) have the overall effect of converting a single molecule of iso-citrate and one of acetyl CoA into one of succinate and one of malate. However, the two enzymes of the bypass are not located in the mitochon-dria, but within cytoplasmic organelles called glyoxysomes, and so its operation as an anaplerotic pathway might involve the operation of tri-carboxylic and dicarboxylic acid transporters (Ch. 4). The glyoxylate cycle bypasses the reactions of the TCA cycle between isocitrate and succinate, i.e. the CO_2 releasing reactions, therefore two carbon atoms are not released as CO_2 when the glyoxylate cycle is operating.

Malate dehydrogenase (decarboxylating), sometimes called malic en-zyme (EC.1.1.1.40) may also act to 'top-up' the TCA cycle, it is a cytosolic enzyme which catalyses the following reaction:

$$Pyruvate + CO_2 + NADPH + H^+ \rightleftharpoons Malate + NADP^+$$

The malate can be fed into the TCA cycle. However, this enzyme probably plays the more important role in adipose tissue of generating NADPH (the reverse reaction) for fatty acid synthesis; it also occurs in heart and muscle. Figure 2.7 shows the pyruvate—malate cycle, which produces one

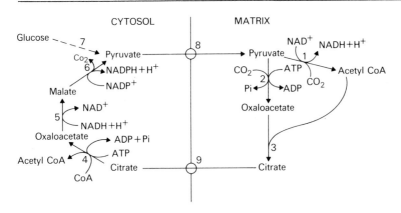

Fig. 2.7   The pyruvate—malate cycle.
1. Pyruvate dehydrogenase complex. 2. Pyruvate carboxylase. 3. Citrate synthase.
4. Citrate lyase. 5. Malate dehydrogenase. 6. 'Malic enzyme'. 7. Glycolysis.
8. Pyruvate transporter. 9. Tricarboxylate transporter.

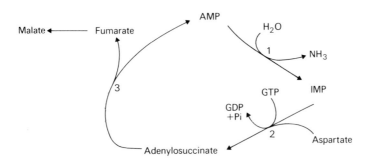

Fig. 2.8   The purine nucleotide cycle.
1. AMP deaminase (EC.3.4.5.6.). 2. Adenylosuccinate synthetase (EC.6.3.4.4.).
3. Adenylosuccinate lyase (EC.4.3.2.2.).

molecule of acetyl CoA and one of NADPH in the cytosol for fatty acid synthesis.

It has also been suggested that the purine nucleotide cycle (Fig. 2.8) may have an anaplerotic role. It produces fumarate in the cytosol which can be converted to malate, which will 'top-up' the TCA cycle. The main role of this cycle is probably to produce ammonia from aspartate in tissues where there is little glutamate dehydrogenase activity.

Coenzyme A

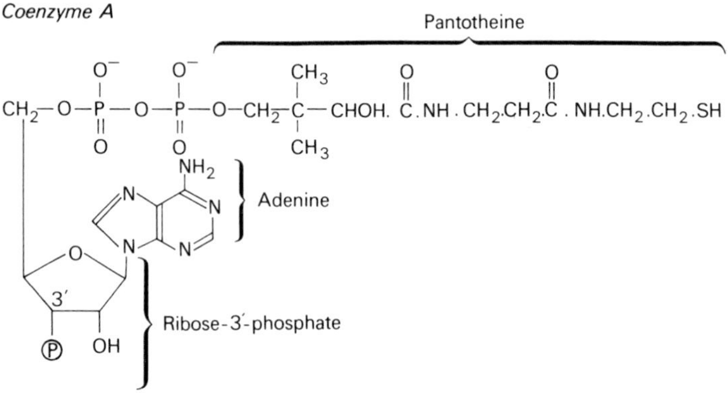

Thiamine Pyrophosphate (TPP)

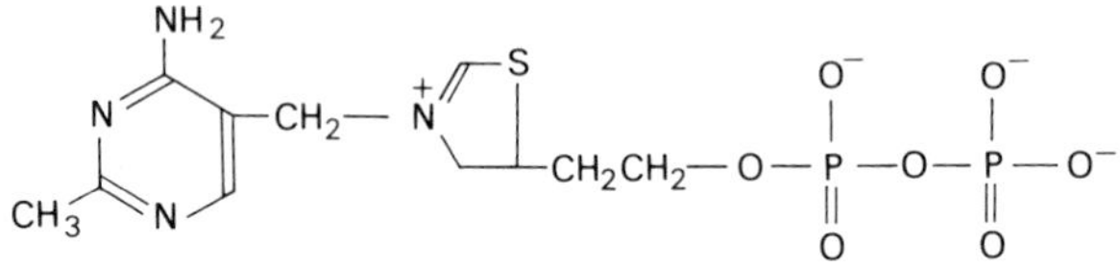

Fig. 2.9   The structure of two cofactors involved in the oxidation of pyruvate.

2.5   Structural organisation of the TCA cycle

With a small number of exceptions the enzymes of the TCA cycle appear to lack any integrated structural organisation and are free in solution in the mitochondrial matrix. Succinate dehydrogenase, however, is firmly bound to the inner membrane of mitochondria. Why this step should be any different from the other steps is unclear. It is the only enzyme which participates in the TCA cycle and in the electron transport system by virtue of the FAD and the non-haem iron protein (iron-sulphur protein) which are an integral part of the enzyme. The involvement of succinate dehydrogenase in the electron transport system is developed in Chapter 3.

The oxidative decarboxylation of pyruvate and 2-oxoglutarate is carried out by the multienzyme pyruvate and 2-oxoglutarate dehydrogenase complexes respectively. These two highly organised multienzyme complexes seem to be very similar in structure and reaction sequence, so only the pyruvate dehydrogenase complex, which has been studied in greater detail, is considered here. The complex is composed of five different enzymes, three of which participate in the reaction sequence, the other two having regulatory roles. There are five cofactors involved; thiamine pyrophosphate (TPP), lipoic acid (LA), coenzyme A, FAD and NAD$^+$. The struc-

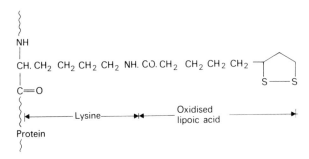

Fig. 2.10   The lipoyllysyl side chain of E_2 (lipoate acetyl transferase).

tures of FAD and NAD^+ are given in section 3.2, lipoic acid in Fig. 2.10 and TPP and CoA in Fig. 2.9.

The three major enzymes are pyruvate decarboxylase *(EC.1.2.4.1), lipoate acetyl transferase (EC.2.3.1.12) and lipoamide dehydrogenase (EC.1.6.4.3). The reaction sequence involved in the conversion of pyruvate to acetyl CoA by these three enzymes is shown in Fig. 2.11. The first step is a nucleophilic attack by the thiazole ring of the pyruvate decarboxylase (E_1)-bound TPP on the pyruvate, resulting in the elimination of CO_2 from pyruvate leaving an hydroxyethyl residue bound to TPP (Fig. 2.12). Next the hydroxyethyl residue is transferred to the lipoic acid bound to lipoate acetyl transferase (E_2). This reaction involves a coupled lipoic acid disulphide reduction and hydroxyethyl to acetyl oxidation (Fig. 2.11). The acetyl residue is then donated to coenzyme A to give the product acetyl CoA; and the lipoic acid is reoxidised by lipoamide dehydrogenase (E_3)-bound FAD. $FADH_2$ is reoxidised by NAD^+, which is reduced to NADH. The key enzyme seems to be the acetyl transferase, which is located at the core of the complex. The overall reaction for the complex is:

$$\text{Pyruvate} + \text{CoA} + NAD^+ \rightleftharpoons \text{Acetyl CoA} + CO_2 + \text{NADH} + H^+$$

In pig heart mitochondria the complex has a molecular weight of about 10^7. It has a number of subunits, there is a core of about 60 lipoate acetyl transferase (E_2) molecules each containing a covalently bound lipoic acid. This core is surrounded by about 24 molecules of pyruvate decarboxylase (E_1 each of M_r 154 000) and 12 molecules of lipoamide dehydrogenase (E_3 each of M_r c. 110 000). Figure 2.13 shows a schematic representation of what the complex may look like, as suggested by electron micrographs.

*Throughout this chapter the name pyruvate decarboxylase will be used for the enzyme whose recommended name is pyruvate dehydrogenase (lipoate), this is in order to distinguish it as being the enzyme which removes CO_2 from pyruvate, i.e. not the enzyme complex.

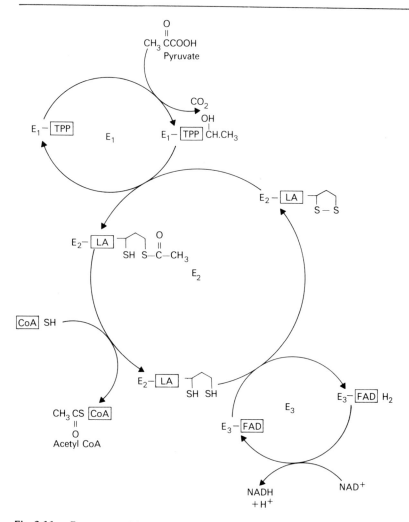

Fig. 2.11 E_1-pyruvate dehydrogenase, E_2-lipoate acetyl transferase, E_3-lipoamide dehydrogenase. LA — lipoic acid.

The exact number of subunits of each enzyme in the complex is in some dispute at present.

Precisely how the complex operates is not clear, but it has been shown that in the *E. coli* complex (and probably in the mitochondrial complex) the five carbon side chain of lipoic acid is joined by a peptide bond to a terminal (ϵ) amino group of a side chain of a lysine residue in the acetyl

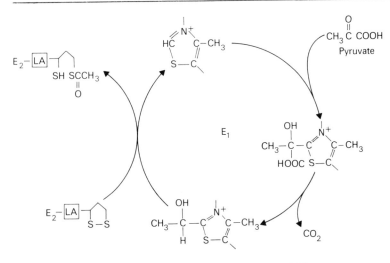

Fig. 2.12   Reactions catalysed by pyruvate decarboxylase (E_1).

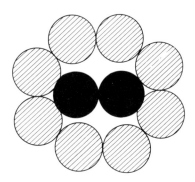

Fig. 2.13   Schematic diagram of the pyruvate dehydrogenase complex. The centre of the complex (white) is the transacetylase enzyme E_2. The ⊘ circles represent the decarboxylase (E_1) and the black circles represent the dehydrogenase (E_3). The complex could have six faces as shown, i.e. each molecule of E_1 shown will be part of another face.

transferase (Fig. 2.10). This means that the dithiolane ring of lipoic acid is attached to a flexible arm approximately 14 Å long. This could be important in permitting the dithiolane ring to flip from the pyruvate dehydrogenase to the lipoamide dehydrogenase.

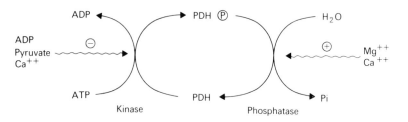

Fig. 2.14 Regulation of PDH by phosphorylation.
$\ominus$ implies inhibition and $\oplus$ implies activation.

The two regulatory enzymes appear to be associated with the pyruvate decarboxylase (E_1) and control the activity of this protein. The first regulatory protein is pyruvate dehydrogenase kinase which catalyses an ATP-dependent phosphorylation of E_1. E_1 consists of two subunits α and β, and the phosphate is added on to the α subunit, resulting in the inactivation of the pyruvate decarboxylase (E_1). Reactivation is achieved by the action of pyruvate dehydrogenase phosphatase, which removes the inhibitory phosphate by hydrolysis. Figure 2.14 shows the interconversion of the phosphorylated and dephosphorylated forms of pyruvate dehydrogenase. Pyruvate dehydrogenase kinase is inhibited by ADP, Ca^{++} and pyruvate and the phosphatase is activated by Mg^{++} and Ca^{++}. Therefore the action of ADP, Ca^{++}, Mg^{++} or pyruvate would bring about the dephosphorylation or activation of the pyruvate dehydrogenase complex. Pyruvate dehydrogenase activity is inhibited by high ratios of $NADH/NAD^+$ or acetyl CoA/CoA, i.e. end-product inhibition. This may be achieved by the activation of pyruvate dehydrogenase kinase.

The 2-oxoglutarate dehydrogenase complex is considered to follow a very similar reaction sequence. The three catalytic enzymes are 2-oxoglutarate decarboxylase (EC.1.2.4.2), lipoate acetyl transferase (EC.2.3.1.12) and lipoamide dehydrogenase (EC.1.6.4.3). The reaction products are succinyl CoA, CO_2 and NADH. The appearance of the isolated complex in electron micrographs is very similar to that of pyruvate dehydrogenase.

2.6   The urea cycle

In mammals excess ammonia is excreted as urea, which is formed by the urea cycle in liver. This cycle (Fig. 2.15) occurs partly in the mitochondrial matrix and partly in the cytosol. There are two amino groups in urea, one is derived from aspartate and the other from free ammonia. Both of these could be derived from glutamate by transamination and the action of glutamate dehydrogenase respectively.

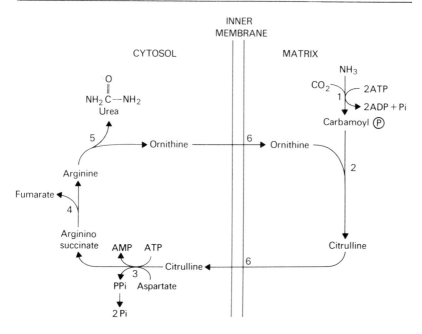

Fig. 2.15 The urea cycle.
1. Carbamoyl phosphate synthetase (EC.2.7.2.5.). 2. Ornithine transcarbamylase (EC.2.1.3.3.). 3. Arginino succinate synthase (EC.6.3.4.5.). 4. Arginino succinate lyase (EC.4.3.2.1.). 5. Arginase (EC.3.5.3.1.). 6. Transporters? (sect. 4.7.2.).

The first reaction of the cycle, which occurs in the matrix, is the formation of carbamoyl phosphate from ATP, CO_2 and ammonia, the enzyme is carbamoyl phosphate synthetase and is activated by N-acetyl glutamate. Ornithine carbamoyl transferase, sometimes called ornithine transcarbamylase, releases the phosphate from carbamoyl phosphate by its reaction with ornithine to give citrulline, this reaction also takes place in the matrix. The citrulline then reacts with aspartate (the second amino group for urea) to give argininosuccinate, which then breaks down to give arginine and fumarate. The enzymes which catalyse the synthesis and breakdown of argininosuccinate are argininosuccinate synthetase and lyase respectively and they are in the cytosol. These four enzymes are all present in animals which can synthesise arginine. The enzyme arginase, which only occurs in ureotelic (urea-producing) animals, catalyses the next reaction, the conversion of arginine into urea and ornithine. The ornithine is recycled since it is transferred into the matrix where it reacts with more carbamoyl phosphate. The activity of arginase can be so rapid that despite having all the enzymes for arginine synthesis, arginine can be an essential amino acid in ureotelic animals as it is all used to produce urea before it can be used for protein synthesis. The overall reaction for the urea cycle is:

$$2NH_3 + 3ATP + CO_2 + 2H_2O \rightleftharpoons 2ADP + 4Pi + AMP + Urea$$

The ammonia and aspartate for urea synthesis are both derived from the breakdown products of proteins, i.e. amino acids. The carbon skeletons of the amino acids can be completely oxidised by the TCA cycle to give ATP. The *net* ATP yields for the breakdown of amino acids can be calculated as were those from the breakdown of carbohydrates (sect. 2.2) and fats (sect. 2.7), by calculating the ATP/NADH used or produced by the urea cycle and the TCA cycle (this will not be calculated here). The end-product of amino acid catabolism is usually glutamate, since amino acids will transaminate with oxoglutarate to give glutamate, or ammonia will react with oxoglutarate to give glutamate (this is the favoured direction of glutamate dehydrogenase). Therefore excess amino acids or ammonia are toxic because they remove oxoglutarate from the TCA cycle causing the cycle to stop. Then NADH (and therefore ATP) will not be produced. The free ammonia for urea synthesis is thought to come from glutamate by the action of glutamate dehydrogenase. However, there is some evidence to suggest that NH_3 may come from the action of the purine nucleotide cycle (Fig. 2.8).

2.7 ß-oxidation of fatty acids

Fat storage occurs largely in specialised adipose tissues. The release of hormones, e.g. adrenaline, causes the hydrolysis of the stored triglycerides in the adipose tissue, which releases free fatty acids and glycerol. The fatty acids are passed to the blood stream where they are transported, bound to serum albumin, to the tissues, muscle and liver, where they are used and undergo oxidation. With the exception of some insect flight muscle, all types of mitochondria are capable of ß-oxidation of fatty acids, which accounts for the bulk of cellular fatty acid oxidation.

The overall process of ß-oxidation is represented in Fig. 2.16. The fatty acids are activated in the cytosol and transported into the mitochondrial matrix as fatty acyl carnitine. This is formed by the sequential operation of fatty acyl thiokinase (on the outer membrane) catalysing reaction 1 and carnitine acyl (palmitoyl) transferase (in the intermembrane space) catalysing reaction 2. An inner membrane-bound carnitine acyl transferase catalyses the reverse of reaction 2, which regenerates fatty acyl CoA in the matrix. The transport of fatty acyl carnitine is discussed further in section 4.8. Fatty acyl CoA can also be generated in the matrix from intramito-

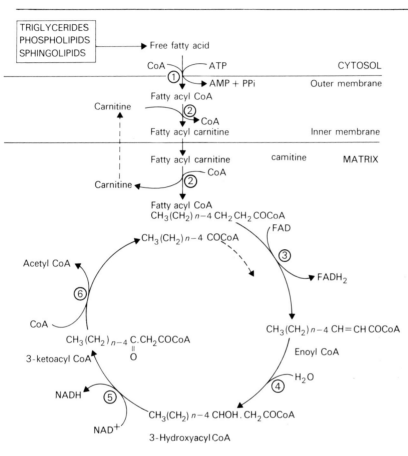

Fig. 2.16 β-Oxidation of fatty acids.
1. Fatty acyl thiokinase (EC.6.2.1.3.). 2. Carnitine acyl transferase (EC.2.3.1.21.).
3. Acyl CoA dehydrogenase (EC.1.3.99.3.). 4. Enoyl CoA hydratase (EC.4.2.1.17.).
5. 3-Hydroxyacyl CoA dehydrogenase (EC.1.1.1.35.). 6. Acetyl CoA: acetyl transferase (EC.2.3.1.16.).

chondrial free fatty acids by the action of one of a number of fatty acid thiokinases each specific for particular chain length fatty acids.

Figure 2.16 shows the β-oxidation of a saturated fatty acid containing n carbon atoms. This is completely oxidised to acetyl CoA ($n/2$ molecules of acetyl CoA per fatty acid molecule) by the repetitive operation of a four-enzyme sequence. The fatty acyl CoA is first dehydrogenated to give enoyl CoA; the hydrogens reduce FAD giving $FADH_2$. The enoyl CoA is then hydrated to give an hydroxyacyl CoA, which is dehydrogenated to

3-ketoacyl CoA, the hydrogens reducing NAD^+ to NADH this time. Then a thiolytic cleavage gives acetyl CoA and a fatty acyl CoA, two carbon atoms shorter than the starting fatty acyl CoA. The enzymes are respectively, acyl CoA dehydrogenase (3), enoyl CoA hydratase (4), 3-hydroxyacyl CoA dehydrogenase (5) and acetyl CoA: acetyl transferase (6). The new fatty acyl CoA can undergo further rounds of β-oxidation by these enzymes until it is completely converted into acetyl CoA. Two enzymes of the sequence (3) and (6) usually occur in multiple molecular forms each specific for different fatty acid chain lengths, short (2–4 carbon atoms), medium (4–12 carbon atoms) and long (13–22 carbon atoms).

The groups arranged across the double bond in the enoyl CoA are in the *trans* configuration, unsaturated fatty acids usually have the *cis* configuration. Therefore unsaturated fatty acids can only be oxidised completely by the β-oxidation enzymes if the configuration about their double bonds is converted from *cis* to *trans*. This is achieved by the enzyme enoyl CoA isomerase.

For ATP generation, quantitatively the most important outcome of β-oxidation of fatty acids is their conversion to acetyl CoA. This can be further oxidised by the TCA cycle with the generation of 11 ATPs and 1 GTP (ATP) per acetyl CoA oxidised. However, a substantial quantity of ATP can also be formed during the operation of the β-oxidation pathway itself. The two dehydrogenase enzymes catalyse coupled reduction of inner membrane cofactors as do the dehydrogenases of the TCA cycle. Fatty acyl CoA dehydrogenase contains both FAD which becomes reduced on the dehydrogenation of fatty acyl CoA. The enzyme reduces a specific inner membrane-bound FAD-containing enzyme, the electron-transferring flavoprotein (ETF). The reduced FAD of the ETF can be reoxidised by oxygen via the electron transport system coupled to the phosphorylation of 2 ADP molecules (Fig. 3.11). The second dehydrogenation step results in the coupled dehydrogenation of 3-hydroxyacyl CoA and the reduction of NAD^+ to NADH. The NADH is reoxidised by oxygen via the electron transport system coupled to the phosphorylation of 3 ADP molecules.

Complete oxidation of a saturated fatty acid of n carbon atoms to CO_2 and H_2O will yield $5(n/2 - 1)$ATP during the β-oxidation process (as $n/2 - 1$ repetitions of the sequence are necessary to give $n/2$ acetyl CoA molecules, and each repetition gives 5 ATP molecules). The $n/2$ acetyl CoA molecules produce $6n$ ATPs when they are oxidised by the TCA cycle (if the GTP is included). This gives a total of $(8.5n - 5)$ ATP molecules per fatty acid molecule. It is necessary to subtract at least 1 ATP to account for that (or the GTP) used in the initial fatty acid activation, leaving $(8.5n - 6)$ ATP molecules. Thus the complete oxidation of palmitic acid $(CH_3(CH_2)_{14}COOH)$ will yield 130 molecules of ATP. Complete oxidation of palmitic acid to CO_2 and water occurs with a free energy change of -9790 kJ mol^{-1} (2340 kcal mol^{-1}) under standard conditions. Phosphorylation of 130 moles of ADP to give ATP requires $130 \times 30.5 = 3965$ kJ (948 kcal). Thus the efficiency of palmitate oxidation is approximately 40 per cent.

The interaction of the fatty acid oxidation pathways with other metabolic sequences occurs mainly through the entry of acetyl CoA into the TCA cycle. From here it can contribute to the carbon skeletons of amino acids. However, because the two-carbon acetyl group is oxidised to $2 CO_2$ by the TCA cycle, animal tissues are incapable of *net* synthesis of carbohydrate from acetyl CoA, there therefore from fatty acids. In plants and microorganisms acetyl CoA can be utilised for carbohydrate synthesis by the sequential operation of the glyoxylate bypass (Fig. 2.6) and phosphoenolpyruvate carboxykinase (Fig. 2.4). The phosphoenolpyruvate so formed can be converted into glucose.

With the exception of those enzymes involved in the extramitochondrial activation of fatty acids and their transport into the mitochondria, the enzymes of the β-oxidation sequence are all in solution in the mitochondrial matrix, and they probably do not form an enzyme complex similar to that for fatty acid synthesis which has been found in the cytosol of *E. coli* and yeast.

2.8 Regulation of mitochondrial metabolism

One of the most fundamental controls of mitochondrial metabolism is that of respiratory control, which is described later (sect. 3.6). In short, there is normally a strict coupling between electron transport and ATP synthesis and if ATP synthesis cannot occur -- because of a low ADP/ATP ratio, this coupling will prevent the oxidation of those cofactors, NADH and $FADH_2$, which initiate the electron transport system. The lack of oxidised cofactors (NAD^+ and FAD) will in turn inhibit those reactions of the TCA cycle and fatty acid oxidation which depend on coupled oxidation-reduction. Clearly this is an important feedback which prevents the unnecessary burning of fuel when ATP supplies are high.

Controlling enzymes of a metabolic pathway usually catalyse non-equilibrium reactions. It has been difficult to determine which of the TCA cycle are non-equilibrium. It is thought that in most tissues and organisms one or more of the following enzymes controls the TCA cycle: (1) pyruvate dehydrogenase (and oxoglutarate dehydrogenase), (2) citrate synthase and (3) NAD^+-linked isocitrate dehydrogenase.

1. *Pyruvate dehydrogenase.* As previously mentioned pyruvate dehydrogenase is controlled by feedback inhibition by high ratios of acetyl CoA/CoA and $NADH/NAD^+$ and by phosphorylation of the decarboxylase enzyme in the complex causing inactivation of the enzyme. The interconversion of the active and inactive forms of pyruvate dehydrogenase is

carried out by a phosphatase and a kinase (Fig. 2.14). In the presence of pyruvate and ADP (which competes with the ATP) the kinase is inhibited thereby preventing the inactivation of the pyruvate dehydrogenase complex. High ratios of NADH/NAD$^+$ and acetyl CoA/CoA may activate the kinase allowing phosphorylation and therefore inactivation of the pyruvate dehydrogenase complex. During starvation fatty acids are used as fuel for energy in preference to carbohydrates. Fatty acids are broken down by β-oxidation to give acetyl CoA, then during starvation pyruvate dehydrogenase will be inhibited, preventing the unnecessary breakdown of carbohydrates.

Ca^{++} inhibits the kinase and activates the phosphatase, therefore Ca^{++} will bring about an activation of pyruvate dehydrogenase. This mechanism will be important in muscle where glycogen is the main source of energy, and Ca^{++} release stimulates muscle contraction.

The oxoglutarate dehydrogenase complex has not been studied as much as that of pyruvate dehydrogenase. Probably increased NADH/NAD$^+$ and acetyl CoA/CoA ratios will inhibit its action and ADP and Ca^{++} will exert similar controls on the complex to those they exert on the pyruvate dehydrogenase complex.

2. *Citrate synthase.* The availability of the substrates for citrate synthase, acetyl CoA and oxaloacetate, will govern the rate of this enzyme. Anaplerotic reactions (sect. 2.4) provide these substrates to allow the TCA cycle to continue. *In vitro* it has been shown that citrate synthase is inhibited by ATP, this would be a sort of feedback inhibition, as the TCA cycle produces reducing equivalents for the production of ATP by oxidative phosphorylation. It has been suggested that in heart, succinyl CoA may control citrate synthase by competing with acetyl CoA. Insect flight muscle mitochondria, however, have such large amounts of citrate synthase in their matrices that even if ATP or succinyl CoA did inhibit its activity it still would not be the rate-limiting enzyme of the TCA cycle.

3. *NAD$^+$-linked isocitrate dehydrogenase.* In some tissues, e.g. insect flight muscle mitochondria NAD$^+$-linked isocitrate dehydrogenase is probably the mmain controlling enzyme of the TCA cycle. In most tissues it is inhibited by ATP and NADH. ADP activates the enzyme (Fig. 2.17) by lowering the K_m for isocitrate. Ca^{++} also has an effect on isocitrate dehydrogenase, at concentrations $>10^{-7}$M, Ca^{++} inhibits the enzyme. This may be an important control of the TCA cycle in heart if Ca^{++} were taken up and released by mitochondria during muscle contraction (sect. 4.9.2).

Pyruvate carboxylase is an important controlling enzyme in mitochondria, it catalyses an anaplerotic reaction (sect. 2.4). When fatty acids are being oxidised, this leads to a high level of acetyl CoA in the matrix and pyruvate dehydrogenase is inhibited, but pyruvate carboxylase will be activated. This allows the conversion of pyruvate to oxaloacetate which

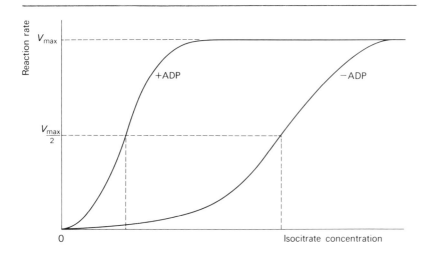

Fig. 2.17   The activation of NAD $^+$-linked isocitrate dehydrogenase by ADP. K_m is the isocitrate concentration when the rate of the reaction is half maximal ($V_{max}/2$). In the presence of ADP the K_m is lower than in the absence of ADP.

can then condense with the acetyl CoA. Figure 2.18 summarises some of the metabolic controls mentioned in this section.

Apart from these metabolic controls on enzyme activity there are other important levels of control operating in mitochondria. The rate of operation of fatty acid oxidation and the TCA cycle can be coarsely regulated by controls on enzyme synthesis. Bakers' yeast (*Saccharomyces cerevisiae*) provides a most striking example of this. In the absence of oxygen and in the presence of high levels of fermentable sugars (glucose, sucrose and fructose) the yeast ceases to synthesise many of the electron chain enzymes (sect. 5.5), this leads to a shut down of normal TCA cycle activity. Instead of oxidising pyruvate to acetyl CoA the yeast converts the pyruvate to ethanol. Precisely why glucose and sucrose repress respiratory enzyme synthesis is not clear – but it is a fortuitous biochemical situation which is much exploited by the brewers and vintners (and drinkers of their products).

This coarse control by the level of enzyme activity can also be seen in the urea cycle in liver. If the animal is fed on a high protein diet the level of some of the enzymes of the urea cycle, notably ornithine transcarbamylase, increases. If the animal is fed on a high carbohydrate diet the levels of some of the enzymes of the urea cycle fall. That is if there is unlikely to be excess nitrogen to be excreted in the form of urea, on the high carbohydrate diet, then the activity of the urea cycle is decreased.

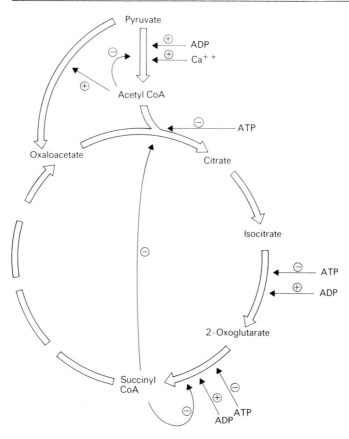

Fig. 2.18   The control of the TCA cycle.
⊖ implies inhibition, ⊕ implies activation.

Other possible points of control of mitochondrial metabolism are at the various steps which require the transfer of metabolites across the mitochondrial membrane by specific carrier systems (Ch. 4). This could be important in controlling the oxidation of fatty acids which have usually to enter the mitochondrial matrix from the cytosol before they can be oxidised by the β-oxidation enzymes.

It is extremely difficult to determine precisely what controls are operating under any given set of circumstances *in vivo*. This is particularly the case in higher animals where the flux through different mitochondrial metabolic pathways varies considerably from tissue to tissue, under various stress conditions and in illness. For example, in mammalian heart muscle mitochondria under normal conditions TCA cycle operation is relatively slow and is maintained largely by acetyl CoA derived from fatty acid oxidation. Within a very short time of the onset of exercise an enhanced TCA cycle activity ensures — presumably as a result of the depletion of ATP and NADH. This increased activity is supported by glycogen breakdown as this is the more locally available substrate and more readily mobilised than fats.

2.9 Other metabolic reactions occurring in mitochondria

Although mitochondria are mainly associated with the TCA cycle, β-oxidation and oxidative phosphorylation, it should be remembered that many other reactions are associated with them, including those associated with mitochondrial nucleic acid and protein synthesis (Ch. 5). In section 1.3.2 and 1.4 some reactions catalysed by enzymes associated with the outer membrane and the intermembrane space are given.

Although it is generally associated with the drug metabolising cycle in endoplasmic reticulum (microsomes) of liver cells, a cytochrome P450 system (see also sect. 3.10) involving a flavoprotein and an iron-sulphur protein also occurs in some mitochondria. In adrenal cortex mitochondria such a system is involved in the hydroxylation of steroids. This is a very important step in the synthesis of steroid hormones.

Some of the enzymes for the biosynthesis of the haem group for cytochromes are present in mitochondria. These include protoporphyrinogenase, an enzyme involved in the synthesis of protoporphyrin IX (sect. 3.2.5), and ferrochelatase (EC.4.99.1.1), the enzyme involved in putting the iron atom into the porphyrin ring.

Mitochondria from liver are associated with the formation of ketone bodies; acetoacetate and 3-hydroxybutyrate are ketone bodies. The production of these in small amounts is a normal process and they are metabolised by peripheral tissues (muscle). They are formed from excess acetyl CoA by the following reactions:

Acetyl CoA + Acetyl CoA $\underset{1}{\rightleftharpoons}$ Acetoacetyl CoA + CoA
Acetoacetyl CoA + Acetyl CoA + H_2O $\underset{2}{\rightleftharpoons}$ β-Hydroxy β-methyl glutaryl CoA + CoA
β-Hydroxy β-methyl glutaryl CoA $\underset{3}{\rightleftharpoons}$ Acetoacetate + Acetyl CoA
Acetoacetate + NADH + H^+ $\underset{4}{\rightleftharpoons}$ 3-Hydroxybutyrate + NAD^+

The enzymes are: (1) acetyl CoA: acetyl transferase (EC.2.3.1.9); (2) HMG CoA synthetase (EC.4.1.3.5); (3) HMG CoA lyase (EC.4.1.3.4); and (4) 3-hydroxybutyrate dehydrogenase (EC.1.1.1.30, see also Table 3.1 and Fig. 3.23) and are all situated in the mitochondria.

A condition called ketosis occurs during starvation and diabetes, when, respectively, carbohydrate stores have been used up or are not metabolised, and then the oxidation of fatty acids is inefficient leading to increased levels of acetyl CoA. The excess acetyl CoA is converted into ketone bodies and large amounts of these occur in the blood and urine. During starvation the brain, which usually uses only glucose as a source of energy can adapt to metabolise ketone bodies.

Suggested further reading

Methods
BERGMEYER, H. U. (1974) *Methods in Enzymatic Analysis* (2nd English edn. 4 volumes). Academic Press, London and New York.

Articles and books
DENTON, R. M. and POGSON, C. I. (1976) *Metabolic Regulation. Outline Studies in Biology Series*, ed. J. M. Ashworth.
FRITZ, I. B. (1964) Carnitine and its role in fatty acid metabolism, *Adv. in Lipid Res.*, 1, 285--334.
GOODWIN, T. W. (1968) ed., *Metabolic Roles of Citrate*. Academic Press.
ISHIKAWA, E., OLIVER, R. M. and REED, L. J. (1966) α-Keto acid dehydrogenase complexes. V. Macromolecular organisation of pyruvate and α-ketoglutarate complexes isolated from beef kidney mitochondria, *Proc. Nat. Acad. Sci.*, 56, 543--41.
KERBEY, A. L., RANDLE, P. J., COOPER, R. H., WHITEHOUSE, S., PASK, H. T. and DENTON, R. M. (1976) Regulation of pyruvate dehydrogenase in rat heart, *Biochem. J.*, 154, 327—48.
KREBS, H. A. (1970) The history of the TCA cycle, *Perspec. Biol. Med.*, 14, 154—70.
LaNOUE, K. F., BRYLA, J. and WILLIAMSON, J. R. (1972) Feedback interactions in the control of citric acid cycle activity in rat heart mitochondria, *J. Biol. Chem.*, 247, 667--79.
LINN, T. C., PELLEY, S. W., PETTIT, F. H., HUCHO, F., RANDALL, D. D. and REED, L. J. (1972) α-Keto acid dehydrogenases. XV. Purification and properties of the component enzymes of the pyruvate dehydrogenase complexes from bovine kidney and heat, *Arch. Biochem. Biophys.*, 148, 327—42.
LOWENSTEIN, J. M. (1969) ed. *Citric Acid Cycle, Control and Compartmentation*. Dekker, New York.
MEHLMAN, M. A. and HANSON, R. W. (1972) ed. *Energy Metabolism and the Regulation of Metabolic Processes in Mitochondria*. Academic Press, New York.
NEWSHOLME, E. A. and START, C. (1974) *Regulation in Metabolism*. John Wiley & Sons, London.

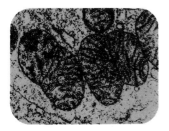

Chapter 3

Oxidative phosphorylation

3.1   Introduction

In the period 1937–41 Kalckar and Belitser independently demonstrated that when various tricarboxylic acid cycle intermediates were oxidised in freshly minced liver there was a concomitant phosphorylation of a number of metabolites such as glucose and fructose. As these metabolites were known to be phosphorylated by the transfer of the terminal phosphate of ATP, this was considered to be evidence that ADP phosphorylation accompanied aerobic oxidations. In the period 1948–50 Kennedy and Lehninger showed that ATP synthesis coupled to oxidation (oxidative phosphorylation) occurs in the mitochondrial fraction of the cells. Since this time the mechanism of ATP synthesis has been one of the most intensively investigated problems in biochemistry.

It is now known that the oxidation process is carried out by an electron transport chain which is situated in the inner mitochondrial membrane. The majority of its components are proteins which have prosthetic groups which are able to undergo oxidation or reduction by the addition or removal of electrons or hydrogen atoms (electrons + protons), the components of the chain being therefore electron or hydrogen carriers. The electrons are supplied to the electron transport chain by the oxidation of metabolites in the mitochondria, e.g. malate to oxaloacetate; succinate to fumarate. Accompanying the electron transport, ATP is synthesised from ADP and inorganic phosphate. This phosphorylation is carried out by the adenosine triphosphatase (ATPase) enzyme complex, which is attached to the inside of the inner membrane. One of the most actively studied problems in this area of biochemistry is how the redox potential energy of

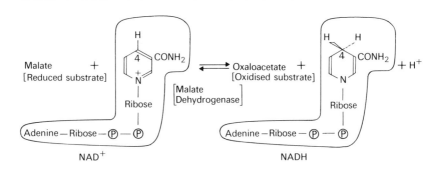

Fig. 3.1   Malate dehydrogenase — catalysed reduction of NAD$^+$ by malate.

the electron transport chain is converted to the 'chemical energy' of ATP. The structure of all the known components of the electron transport chain and ATPase and how these are fuctionally and structurally organised in the mitochondria will be discussed in this chapter.

3.2   Electron carriers

3.2.1 Pyridine-nucleotide-linked dehydrogenases
These are enzymes from the class oxidoreductase which have NAD$^+$ (or NADP$^+$) as coenzymes. This means that NAD$^+$ is involved in the enzyme reaction but is able to freely dissociate from the enzyme. The structures of NAD$^+$ and NADH are shown in Fig. 3.1. NADP has an additional phosphate group on the 2' position of the adenine-linked ribose. The active part of NAD$^+$ is the nicotinamide (or pyridine) residue which can accept another hydrogen at position 4. (The nicotinamide group is the vitamin niacin.) Several pyridine-nucleotide-linked dehydrogenases are associated with the TCA cycle, e.g. malate dehydrogenase and isocitrate dehydrogenase (Ch. 2). Figure 3.1 shows the reaction catalysed by malate dehydrogenase; two hydrogen atoms are removed from malate to form oxaloacetate, of these a proton and two electrons (i.e. a hydride ion, H$^-$) are donated to the nicotinamide group forming NADH, and the other proton is released into the medium. The hydrogen is added stereospecifically to the nicotinamide residue. Malate dehydrogenase adds H from the above plane of the nicotinamide ring (A-side). Some of the other pyridine-nucleotide-linked dehydrogenases cause H to be added from below the plane of the ring (B-side). The NADH formed is a reducing agent which can, in turn donate a hydride ion to another electron carrier, regenerating NAD$^+$.

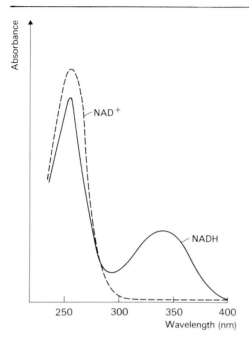

Fig. 3.2   Absorption spectra of NAD⁺ and NADH.

Both NAD^+ and NADH absorb ultraviolet light at 260 nm; this is due to the purine ring structure. The nicotinamide residue in NAD^+ does not absorb long wavelength ultraviolet light, but the addition of H^- to form NADH gives rise to an absorption peak at 340 nm (Fig. 3.2). The 340 nm absorption can be used to measure enzyme reactions catalysed by pyridine-nucleotide-linked dehydrogenases or other enzyme reactions which can be linked to these dehydrogenases. The disappearance or appearance of NADH is followed by measuring adsorption at 340 nm. Alternatively NAD^+ reduction may be followed fluorimetrically by following the increase in fluorescence at 480 nm (illumination at 365 nm) or by following the decrease in pH resulting from the release of a proton during the reduction.

The pyridine-nucleotide-linked dehydrogenases discussed here are not bound to the inner mitochondrial membrane; as previously stated some are involved in the TCA cycle. However, they do provide the reducing equivalents (i.e. electrons), for the electron transport chain in the mitochondrial inner membrane. Table 3.1 lists some of the more important pyridine-nucleotide-linked dehydrogenases involved in this function.

Table 3.1 Mitochondrial pyridine-nucleotide-linked dehydrogenases

Enzyme	EC. number	Function
Isocitrate dehydrogenase (NAD$^+$)	1.1.1.41	TCA cycle enzyme
Malate dehydrogenase	1.1.1.38	TCA cycle enzyme
Dihydrolipoyl dehydrogenase	1.6.4.3	Component of the pyruvate and oxoglutarate dehydrogenase complexes
3-Hydroxyacyl CoA dehydrogenase	1.1.1.35	β-oxidation enzyme
Glutamate dehydrogenase (NAD$^+$)	1.4.1.2	Oxidative deamination of glutamate
3-Hydroxybutyrate dehydrogenase	1.1.1.30	Oxidation of ketone bodies

3.2.2 Flavoproteins

Flavoproteins are also enzymes of the class oxidoreductase. These have either flavin mononucleotide (FMN) or flavin adenine dinucleotide (FAD) as a prosthetic group. There are four main flavoproteins all bound to the inner mitochondrial membrane and involved in the electron transport chain. The most important of these is probably NADH dehydrogenase (EC.1.6.99.3) which contains FMN. The other three, succinate dehydrogenase (EC.1.3.99.1), glycerol-3-phosphate dehydrogenase (EC.1.1.99.5) and electron transferring flavoprotein (ETF) all have FAD as prosthetic groups. With the exception of ETF these flavoproteins contain non-haem iron and, as with other iron-sulphur proteins (see below) the iron is associated with acid-labile sulphur (yields H_2S when treated with acid).

FAD and FMN in the flavin-linked dehydrogenases, unlike NAD$^+$ in pyridine-nucleotide-linked dehydrogenases are tightly bound to the protein part of the enzyme. In the case of succinate dehydrogenase the FAD is covalently linked to a histidine residue in the protein.

The structures of FAD and FMN are shown in Fig. 3.3; oxidation-reduction occurs in the isoalloxazine ring system which is capable of accepting two hydrogens $(2H^+ + 2e^-)$ from the substrate. For example succinate dehydrogenase catalyses the transfer of hydrogens from succinate to FAD yielding $FADH_2$ and fumarate.

FAD is derived from Vitamin B_2 (riboflavin) and like NAD$^+$ its oxidation-reduction can be monitored spectrophotometically as FAD has an absorption maximum of 450 nm which disappears on reduction to $FADH_2$ (Fig. 3.4).

3.2.3 Iron-sulphur proteins

The presence in some mitochondrial flavoproteins of iron associated with acid-labile sulphur has already been mentioned. Including these there are ten iron-sulphur centres in mammalian mitochondria which are detectable by electron spin resonance spectroscopy (ESR) and are involved in elec-

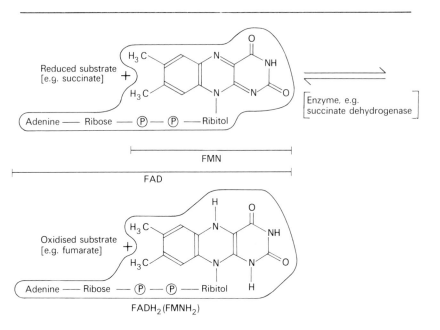

Fig. 3.3   Enzyme reduction of FAD or FMN.

tron transport. They all contain iron atoms covalently bound to the proteins by cysteine sulphurs and to other iron atoms by sulphur bridges (Fig. 3.5). The sulphur bridges are the acid-labile sulphurs. There seem to be about 10 iron-sulphur centres associated with mitochondria; $(Fe\ S)_{1a}$, $(Fe\ S)_{1b}$, $(Fe\ S)_2$, $(Fe\ S)_3$ and $(Fe\ S)_4$ are associated with NADH dehydrogenase. $(Fe\ S)_5$, $(Fe\ S)_6$ and $(Fe\ S)_9$ are associated with Complex III (sect. 3.4) and $(Fe\ S)_7$ and $(Fe\ S)_8$ are associated with succinate dehydrogenase. Some of these centres seem to be of the 4 Fe; 4 acid-labile S variety (Fig. 3.5b) but others may contain 2 Fe; 2 acid-labile S (Fig. 3.5a). Iron-sulphur centres act as electron carriers by undergoing oxidation-reduction between Fe (II) and Fe (III), however, each centre will act as only a single electron carrier despite the presence of more than one Fe. The reduced Fe (II) form contains unpaired electrons which give rise to a low-temperature electron spin resonance spectrum with signals between 1.90 and 2.10 gauss, and this characteristic has been used to study the involvement of iron-sulphur proteins in electron transport. Although they are important as electron carriers, the precise function, and the necessity for so many iron-sulphur centres in the electron transport chain is unknown.

3.2.4 Ubiquinone
Ubiquinone (or coenzyme Q) is a lipid-soluble hydrogen carrier in the

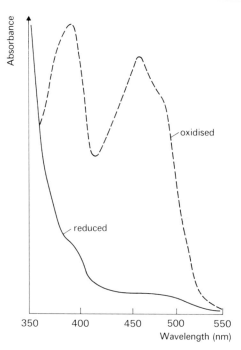

Fig. 3.4   Absorption spectra of oxidised and reduced forms of a flavoprotein.

inner mitochondrial membrane. It is a substituted benzoquinone (Fig. 3.6) with a polyisoprene side chain. The length of the side chain can vary from 0 to 10 isoprene residues (6–10 in naturally occurring ubiquinones). Ubiquinone-(10) (which has 10 isoprene residues) is found in the mitochondria of higher organisms and this long side chain ensures ubiquinone solubility in the hydrophobic interior of the inner membrane. Other members of the ubiquinone family are found in microorganisms. Ubiquinones-(0) and -(1) are often used in experimental studies *in vitro* instead of ubiquinone-(10) because they are water-soluble but still act, to a considerable extent, in the same way as ubiquinone-(10).

Oxidation and reduction occur in the benzoquinone nucleus of the molecule giving the quinone (oxidised) and hydroquinone (reduced) forms (Fig. 3.6) which means that ubiquinone is a hydrogen (proton + electron) carrier. Ubiquinone has an absorption maximum at 275 nm which disappears when it is reduced to UQH_2. This absorption change can be used to study the kinetics of the oxidation and reduction of ubiquinone (Fig. 3.7).

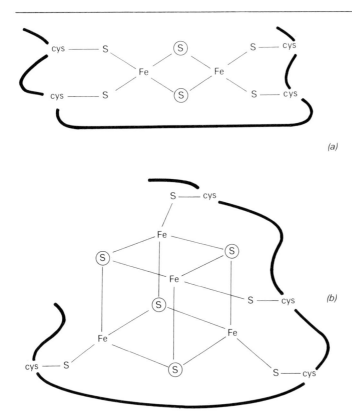

Fig. 3.5 Two types of iron-sulphur centre. (*a*) 2-Fe centre; (*b*) 4-Fe centre. The ringed sulphurs are acid-labile. The thick line represents the covalent backbone of the protein.

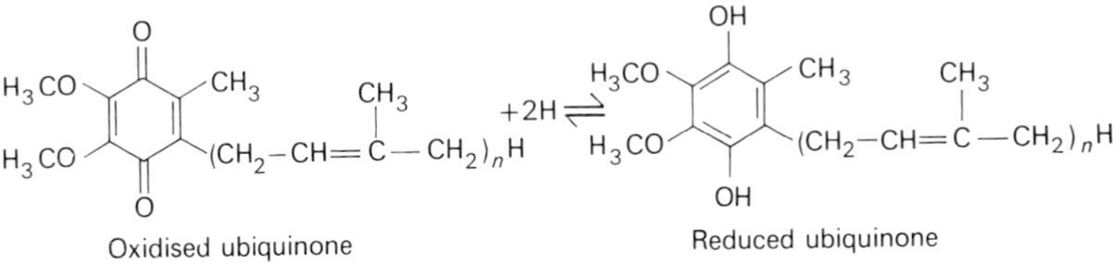

Fig. 3.6 Reduction of ubiquinone.

Oxidised ubiquinone   Reduced ubiquinone

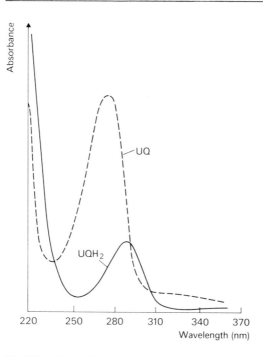

Fig. 3.7   Absorption spectra of oxidised and reduced ubiquinone.

3.2.5 Cytochromes

Cytochromes are proteins containing a prosthetic group called haem. The basis of the haem is porphin which consists of four pyrrole rings linked in a cyclic manner by methene bridges. The porphin nucleus has various groups attached to it to give porphyrins. The most common porphyrin in biological systems is protoporphyrin IX (Fig. 3.8) which has methyl groups in positions 1, 3, 5 and 8, vinyl groups in positions 2 and 4 and propionic acid groups in positions 6 and 7. Haem is a porphyrin ring in which the four pyrrole nitrogens are coordinated to an iron atom forming a square planar complex.

Iron has six coordination positions and usually positions 5 and 6 which are perpendicular to the plane of the porphyrin ring are occupied by side chains of specific amino acids from the protein, e.g. a histidine imidazole group or a methionine sulphydryl group. The sixth coordination position of some cytochromes (e.g. cytochrome a_3 and cytochrome P450) is free to bind O_2 (or carbon monoxide or cyanide). Cytochrome a_3 is the last electron carrier in the electron transport chain and is involved in the

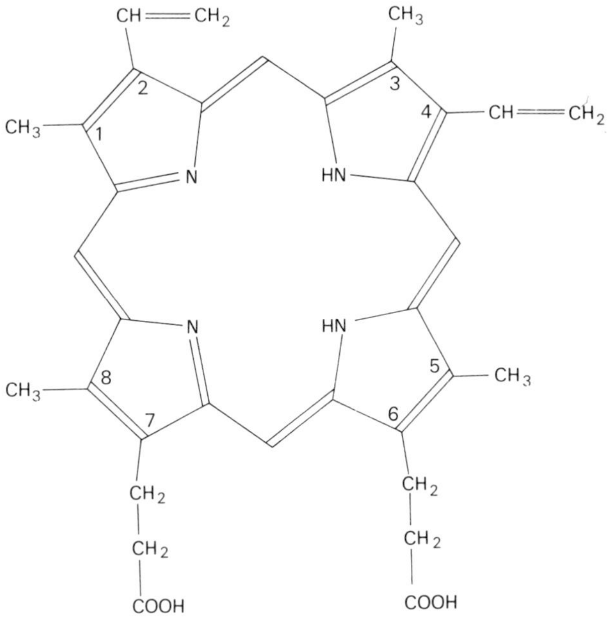

Fig. 3.8   Structure of protoporphyrin IX.

reduction of O_2; cytochrome P450 (so-called because of the sharp peak at 450 nm when it reacts with carbon monoxide) plays a part in steroid and drug hydroxylation in which O_2 is also involved. Cytochromes are classified into three major classes, A, B and C according to the substituents bound to the porphin ring. Table 3.2 lists these substituents and also the particular mitochondrial cytochromes which fall into each class. Class A cytochromes represented in mitochondria by cytochromes a and a_3 contain Haem A, Class B cytochromes — cytochromes b-562 and b-566 (also referred to as b_K and b_T) — contain protohaem and Class C cytochromes (cytochromes c and c_1) contain Haem C. Class C cytochromes are the only ones whose haem is covalently bound to the protein (by positions 2 and 4 of the porphyrin ring). In all of the others the haem is held in position by coordination bonds to the iron atom. All of the cytochromes of the electron transport chain are in the inner mitochondrial membrane and all except cytochrome c, which is loosely bound to the outside (C-side) of the membrane and readily water-soluble, are very difficult to purify. The amino acid sequence of cytochrome c has been established for a large

Table 3.2 The side chains of the haems of mitochondrial cytochromes
(other substituents are as shown in Fig. 3.8)

Cytochrome	Haem	Position of side chain			
		2	4	5	8
aa_3	Haem A	$\overset{\text{CH}_3}{\underset{\text{O}\diagdown\text{C}_{11}\text{H}_{19}}{}}$	$-CH = CH_2$	$-H$	$-CHO$
b	Protohaem	$-CH = CH_2$	$-CH = CH_2$	$-CH_3$	$-CH_3$
c	Haem C	$-CH - CH_3$ | S | protein	$-CH - CH_3$ | S | protein	$-CH_3$	$-CH_3$

number of species and the three-dimensional structures of both oxidised
and reduced cytochrome *c* have been determined by X-ray diffraction
analysis.

The role of cytochromes as electron carriers was first suggested by
Keilin who observed spectroscopically their oxidation and reduction in
insect flight muscle in response to changes in physiological conditions. The
cytochromes undergo oxidation and reduction by a valency change from
Fe (II) (reduced) to Fe (III) oxidised.

$$Cyt.Fe^{++} \rightleftharpoons Cyt.Fe^{+++} + e^-$$

The oxidised forms of most cytochromes have two absorption bands in the
visible wavelength range, while reduced forms have three. Those for
sodium dithionite-reduced cytochrome *c* are shown in Fig. 3.9. The three
peaks are referred to as α, β and γ (or Soret) peaks in descending order of
wavelength.

When studying the cytochrome composition of mitochondria or cells
the production of normal absorption spectra is hampered by the high level
of light scattering which cuts down the transmittance of light through the
sample. Consequently it is usual to measure a difference spectrum; the
sample cuvette contains reduced cells, the reference cuvette an equal
quantity of oxidised cells. In this way the effects of light scattering are
equalised in the two cuvettes. The reduced minus oxidised difference
spectra of cytochrome *c*, for insect flight muscle mitochondria and for the
yeast *Kluyveromyces lactis* are shown in Fig. 3.10. The two latter show α
absorption peaks for Class A, B and C cytochromes at approximately 605,
560 and 550 nm respectively. These were obtained at room temperature;
when measured at liquid N_2 temperature (77 K) the peaks for different
cytochromes within each class can be observed as these are very much
sharpened by the narrower distribution of vibrational and rotational sub-
states at this temperature. The peak maxima for the mitochondrial cyto-

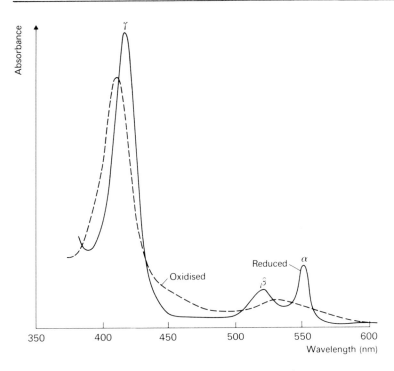

Fig. 3.9 Absorption spectra of oxidised and reduced cytochrome *c*.

chromes are shown in Table 3.3; cytochrome *a* and a_3 do not have a β peak. Generally the α bands are the most characteristic for individual cytochromes and absorption changes of the α absorption bands are used to study oxidation-reduction reactions of specific cytochromes in a mitochondrial preparation. Light scattering is also a problem in this kind of

Table 3.3 Absorption maxima for reduced mammalian cytochromes

Cytochromes	Peaks (nm) α	β	γ
a	600	—	439
a_3	603.5	—	443
b-562 (b_K)	562	532	429
b-566 (b_T)	566	*	*
	+588 *shoulder*		
c	550	521	415
c_1	554	524	418

*Not known.

57

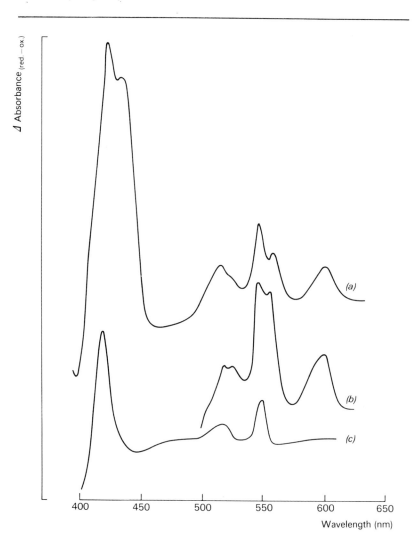

Fig. 3.10 Reduced minus oxidised difference spectra.
(a) Insect flight muscle mitochondria.
(b) Yeast cells (*Kluyveromyces lactis*, courtesy of B. M. Allmark).
(c) Cytochrome c.

investigation as changes in metabolic states of mitochondria result in swelling or contraction of mitochondria which considerably affect light scattering; consequently, using conventional spectrophotometry, it is not possible to distinguish changes in absorbance due to changes in oxidation-reduction of the cytochrome from those due to light scattering. To overcome this problem Chance introduced a technique known as dual-wavelength spectrophotometry. In this a single cuvette is used and two light beams of different set wavelengths are alternately passed through the cuvette. The difference in absorbance between the two wavelengths is recorded continuously. The first wavelength is usually at the peak of the reduced-oxidised difference spectrum for the cytochrome being studied; the second (reference) wavelength is fixed at a point where no changes due to cytochrome oxidation-reduction occur (isosbestic point). As long as the two wavelengths are not too far apart changes in light scattering will be more or less identical at the two wavelengths and so changes in absorbance difference will be due solely to changes in light absorption. Keilin initially postulated the existence of three cytochromes a, b and c in mitochondria. Cytochrome c_1 was discovered when its α peak was observed as a shoulder on the cytochrome c α peak. Cytochrome b occurs in two forms cytochrome b-562 and b-566 which have differing oxidation-reduction potentials (sect. 3.3) and absorption peaks. It is possible that these two cytochromes are identical in structure but have different locations in the membrane. Cytochrome a_3 was discovered when it was found that the cytochrome $a\alpha$ peak was split into two components on treatment with carbon monoxide. The name cytochrome a was retained for the component which does not react with CO. It is suggested that, since cytochromes a and a_3 cannot be separated, cytochrome oxidase, the enzyme complex which catalyses the oxidation of cytochrome c by O_2, is a lipoprotein complex containing two similar haem groups which have different oxidation-reduction kinetics; thus cytochrome c reacts with cytochrome a faster than with cytochrome a_3, O_2 reacts with cytochrome a_3 faster than with cytochrome a and cytochrome a_3 is inhibited by cyanide and carbon monoxide. The complex, which in yeast mitochondria is now known to contain two haem groups, two copper atoms and seven polypeptide subunits is generally referred to as cytochrome aa_3 (sect. 5.5).

3.2.6 Copper

Cytochrome aa_3 contains two copper atoms associated with cytochrome a and one with cytochrome a_3. These coppers are referred to as Cu_A and Cu_B respectively; changes in their ESR signal have been attributed to their undergoing oxidation and reduction between Cu (I) and Cu (II) states. It is thought that electron transport through the cytochrome oxidase complex is initiated by reduction of cytochrome a followed by Cu_A, Cu_B and then cytochrome a_3. It is possible that reduction of Cu_B^{++} to Cu_B^{+} may cause a shift in one of the polypeptide chains to reveal coordination position 6 in the haem of cytochrome a_3, thus allowing oxygen to come in and be reduced.

3.3 Oxidation-reduction (redox) potentials

Any oxidation reaction must be accompanied by a reduction reaction:

$$AH_2 \quad \diagdown \diagup \quad B$$
$$A \quad \diagup \diagdown \quad BH_2$$

So if AH_2 is oxidised to A then simultaneously B will be reduced to BH_2. AH_2/A and B/BH_2 are redox couples. Earlier in this chapter we came across examples of redox couples: malate/oxaloacetate, NADH/NAD⁺, $FADH_2/FAD$. In the reaction catalysed by malate dehydrogenase (Fig. 3.1) malate is oxidised to oxaloacetate while NAD⁺ is reduced to NADH. Similarly in the succinate dehydrogenase reaction succinate is oxidised to fumarate while FAD is reduced to $FADH_2$ (Fig. 3.3).

The oxidation-reduction potential of a couple is a measure of the oxidising or reducing power of that redox couple. A redox couple has a tendency to oxidise (accept electrons from) others of more negative potential and to reduce (donate electrons to) those of more positive potential. The standard oxidation-reduction potential is a measure of the ability of a redox couple to donate electrons or to gain electrons under standard conditions. This is written as E° and is measured at 0 pH and with unit activity of all components. Those couples with an E° less than 0 will tend to reduce a standard hydrogen electrode (H_2 gas at 1 atm pressure in equilibrium with H^+ ions in solution of unit activity).

$$H^+ + e^- \rightleftharpoons \frac{1}{2}H_2$$

In practice redox potentials of respiratory carriers operating *in situ* cannot be measured under the above standard conditions and instead a midpoint potential E'_0 is usually recorded with the system at either 25° or 30°C and

Table 3.4 Midpoint potentials of some redox couples at pH 7.0 and 25°C. The values for the respiratory carriers are measured in isolated mitochondria

Reaction	E'_0 (Volts)
2-Oxoglutarate + CO_2 + 2 H^+ + 2 $e^- \rightleftharpoons$ Isocitrate	−0.380
NAD^+ + 2 H^+ + 2 $e^- \rightleftharpoons NADH + H^+$	−0.320
Fumarate + 2 H^+ + 2 $e^- \rightleftharpoons$ Succinate	−0.031
2 cyt.b_T(ox) + 2 $e^- \rightleftharpoons$ 2 cyt.b_T(red)	−0.030
2 cyt.b_K(ox) + 2 $e^- \rightleftharpoons$ 2 cyt.b_K(red)	+ 0.065
Ubiquinone + 2 H^+ + 2 $e^- \rightleftharpoons$ Ubiquinol	+ 0.100
2 cyt.b_T(ox) + 2 $e^- \rightleftharpoons$ 2 cyt.b_T(red) (in the presence of ATP)	+ 0.240
2 cyt.c_1(ox) + 2 $e^- \rightleftharpoons$ 2 cyt.c_1(red)	+ 0.220
2 cyt.c(ox) + 2 $e^- \rightleftharpoons$ 2 cyt.c(red)	+ 0.254
2 cyt.a_3(ox) + 2 $e^- \rightleftharpoons$ 2 cyt.a_3(red)	+ 0.385
$\frac{1}{2}O_2$ + 2 H^+ + 2 $e^- \rightleftharpoons H_2O$	+ 0.816
2 H^+ + 2 $e^- \rightleftharpoons H_2$	−0.421

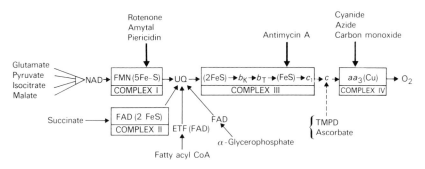

Fig. 3.11   Functional organisation of the electron transport chain.

at pH 7.0 and with concentration of the oxidant and reductant being equal. With E'_0 determinations it is always necessary to stipulate the conditions under which the midpoint potential was measured. Table 3.4 gives midpoint potentials for a number of redox couples associated with the respiratory chain. Determination of these redox potentials has helped to establish the order of the components of the respiratory chain.

3.4   Structural organisation of the electron transport system

Figure 3.11 shows a currently acceptable arrangement of the components of the respiratory chain from NADH to oxygen. This linear sequence allows one component to transfer hydrogens (or electrons) to the next component thereby causing its own oxidation and reduction of the next component in the sequence. A great deal of research has been carried out

Table 3.5   The reduction of components of the respiratory chain under various conditions. The additions of substrate and inhibitors were made to isolated mitochondria in medium

State of mitochondria	Per cent reduction of respiratory chain components				
	NAD^+	Fp	cyt. b	cyt. c	cyt. a
Resting (State IV)	99	40	35	14	0
Active (State III)	53	20	16	6	4
Anaerobic + substrate	100	100	100	100	100
Aerobic — substrate	0	0	0	0	0
Substrate + rotenone	100	100	0	0	0
Substrate + antimycin	100	100	100	0	0
Substrate + KCN	100	100	100	100	100

to determine this order. This has been considerably helped by the fact that the redox states of all of the components of the respiratory chain can be determined spectroscopically. Some of the results which have helped to elucidate the order are summarised in Table 3.5.

Chance and Williams found that if isolated mitochondria were maintained in an anaerobic state with an excess of oxidisable substrate then the respiratory chain components all become completely reduced; but if the system was suddenly oxygenated then cytochrome aa_3 became oxidised first, then cytochrome c then cytochrome b and finally NADH. Obviously the carriers acting most closely to O_2 would be expected to be oxidised first. The order suggested by this experiment was supported by the complementary experiment in which mitochondria were initially kept aerobically without substrate. Then all the components were oxidised. Addition of substrate caused reduction in the order NAD^+, flavoprotein, cytochrome b, cytochrome c and cytochrome aa_3. If mitochondria are maintained in the presence of substrate and oxygen then each carrier achieves a steady-state in which a proportion of the carrier is in the oxidised and a proportion in the reduced form. Those acting closest to the substrate would be expected to have a higher percentage reduction than those acting closer to O_2. The results outlined in Table 3.5 for mitochondria respiring in the presence of ADP (State III respiration) or in its absence (State IV respiration) confirm the results of the other experiments. The order deduced in these and subsequent experiments is summarised in Fig. 3.11. This agrees well with an increasing midpoint potential of the carriers going from substrate to oxygen.

Green, Hatefi and their co-workers fragmented the respiratory chain into four functional complexes plus ubiquinone and cytochrome c by treatment with detergents and ammonium sulphate fractionation. The respiratory chain components of each are indicated in Fig. 3.11. Each complex also contains phospholipid. These complexes numbered I to IV catalyse the following reactions:

I   NADH oxidation — ubiquinone reduction
II Succinate oxidation — ubiquinone reduction
III Reduced ubiquinone oxidation — cytochrome c reduction
IV Cytochrome c (Fe^{++}) oxidation — oxygen reduction.

It has been shown that ubiquinone can transfer reducing equivalents between complexes I and III and between II and III and that cytochrome c can transfer reducing equivalents between III and IV. This work demonstrated a structural basis for the previously established functional organisation. As well as considering the linear relationship between the different carriers it is necessary to examine their three-dimensional spatial arrangement within the inner membrane of the mitochondrion. A great deal of work by Racker and his co-workers in the 1960s and by others subse-

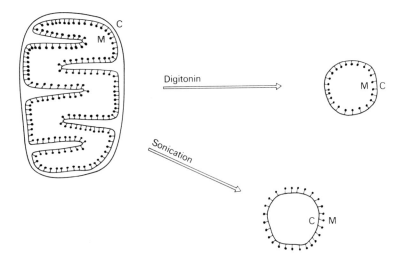

Fig. 3.12   Preparation of submitochondrial particles with normal or inverted membrane orientation.

quently has been carried out on this aspect of the respiratory chain. As described in Chapter 1 the inside of the mitochondrial membrane has spheres attached containing the mitochondrial ATPase activity (see sect. 3.8). This side of the membrane is called the 'M' side (matrix-facing side) while the other side is called the 'C' side (cytochrome c is located here). To investigate the position of the various inner membrane proteins isolated mitochondria can be treated with the detergent digitonin, which removes the outer mitochondrial membrane leaving submitochondrial particles with the 'C' side facing outwards (as in intact mitochondria). If, on the other hand, mitochondria are disrupted by sonication submitochondrial particles are formed, when the cristae fragment, with the 'M' side facing outwards (Fig. 3.12). These two types of submitochondrial particles, which are both capable of carrying out electron transport reactions, have been used to locate the various components of the oxidative phosphorylation system.

One approach has been to raise an antibody against a purified mitochondrial component and to test this antibody for inhibition of electron transport in the two types of particle. If the component is located on the 'C' side of the inner membrane then the antibody can be expected to inhibit electron transport in intact mitochondria or digitonin-prepared submitochondrial particles. As the antibody will be unable to cross the inner membrane it will not inhibit in sonicated submitochondrial particles because of inversion of the normal configuration of the mitochondrial

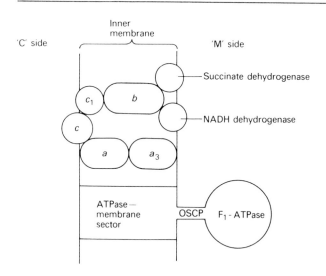

Fig. 3.13   Structural organisation of the oxidative phosphorylation system.

membrane. In contrast a component located on the 'M' side will react with
its antibody only in sonicated submitochondrial particles. A second
approach is to label any proteins on the outwardly-facing surface of the
different submitochondrial particles by reaction with an impermeable
protein reagent such as [^{35}S]-diamino benzenesulphonic acid (DABS) or
by iodination with lactoperoxidase. Subsequent purification of individual
components shows which are labelled with DABS or iodine and were
consequently on the outside surface in the submitochondrial particles in
question.

All of these techniques have shown that cytochrome c is located on the
'C' side a conclusion that is supported by the relative ease with which
cytochrome c is removed from intact mitochondria but not from sonicated
submitochondrial particles. Cytochrome c_1 is also located on the 'C' side
from immunological studies but its greater hydrophobicity and firmer
binding to the membrane suggest a partial immersion within the membrane. Cytochrome b seems to be rather inaccessible to surface acting
agents and is extremely difficult to extract from mitochondria — observations that have led to the suggestion that it is submerged in the hydrophobic central region of the membrane. Cytochrome oxidase (cytochrome
aa_3) gave results which suggested that this enzyme complex straddled the
membrane, being accessible to both surfaces. Furthermore, azide, which
reacts with cytochrome a_3 and only slowly permeates the inner membrane,
is more effective in inhibition of sonicated particles than intact mitochon-

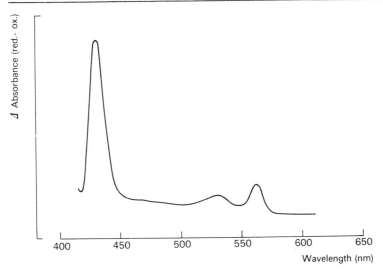

Fig. 3.14   Crossover spectrum.
Difference spectrum of flight muscle mitochondria, where the reduced sample was reduced by pyruvate in the presence of antimycin A. Cytochrome *b* is reduced, cytochromes *a* and *c* are oxidised (compare with Fig. 3.10*a*).

dria. On the other hand polylysine, which reacts with cytochrome *a* reacts only with intact mitochondria or digitonin particles. Isolation of the different cytochrome oxidase polypeptide subunits of cytochrome oxidase from DABS-labelled submitochondrial particles has furthermore shown that some are located at the 'M' side and others at the 'C' side.

Succinate and NADH-dehydrogenase are both lcoated at the 'M' side which is what one would expect from components reacting with TCA-cycle intermediates. Succinate will reduce ferricyanide (which is impermeant) in sonicated particles but not intact mitochondria and NADH (also impermeant) is oxidised in sonicated particles but not intact mitochondria.

The mitochondrial F_1-ATPase (sect. 3.7) has been shown by immunological studies, DABS labelling and electron microscopy to be located on the 'M' side of the inner membrane. These results are summarised diagrammatically in Fig. 3.13.

3.5   Inhibition of the respiratory chain

The respiratory chain can be inhibited by various compounds and this inhibition has yielded valuable information about the order of the components. When an inhibitor is added in the presence of oxygen and an

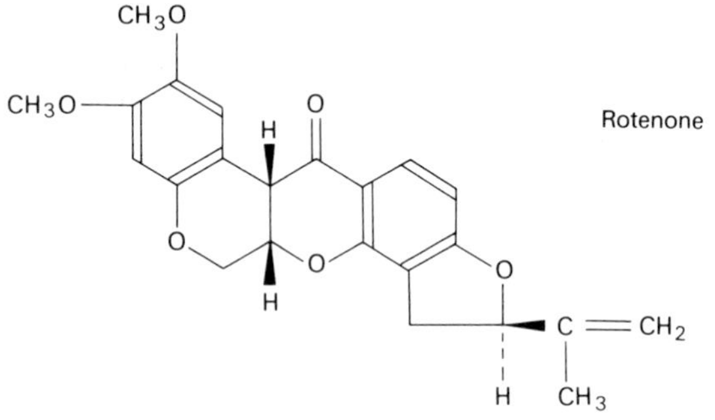

Rotenone

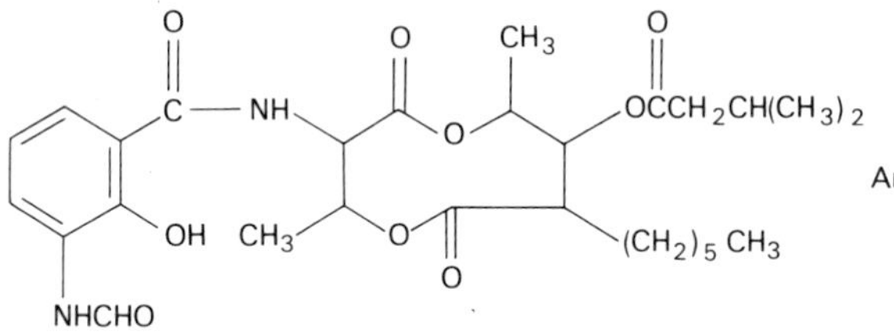

Antimycin A$_1$

Fig. 3.15   Structures of respiratory inhibitors rotenone and antimycin A.

oxidisable substrate all the electron carriers on the oxygen side of the site of action of the inhibitor will become oxidised and all those on the substrate side will be reduced. Chance has referred to this phenomenon as a crossover point and these are shown in Table 3.5 for the inhibitors rotenone, antimycin A and cyanide and in Fig. 3.14 for antimycin A. Rotenone, amytal and piericidin inhibit between NADH dehydrogenase and ubiquinone (i.e. in Complex I), antimycin A inhibits between cytochrome b and cytochrome c_1 (Complex III) and cyanide, carbon monoxide and azide inhibit cytochrome oxidase (Complex IV). The structures of rotenone and antimycin A are shown in Fig. 3.15 and the sites of action of a number of inhibitors are indicated in Fig. 3.11.

So far in this chapter we have only discussed the electron transport involved in oxidative phosphorylation. Electron transport is coupled to the phosphorylation of ADP which is carried out by the ATPase enzyme complex. ATPase in intact mitochondria catalyses the synthesis of ATP and is sometimes referred to as ATP synthetase; if ATPase is removed from the membrane it will catalyse the hydrolysis of ATP to ADP and inorganic phosphate. Whatever the direction of enzyme action in mitochondria, it

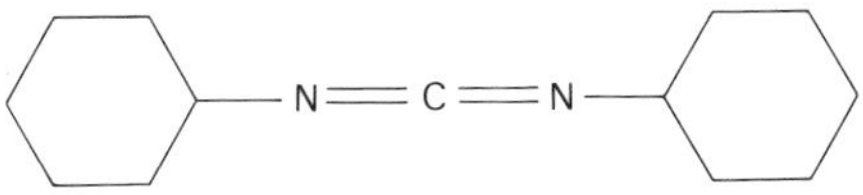

Dicyclohexylcarbodiimide (DCCD)

Fig. 3.16 The structure of the ATP synthesis inhibitor DCCD.

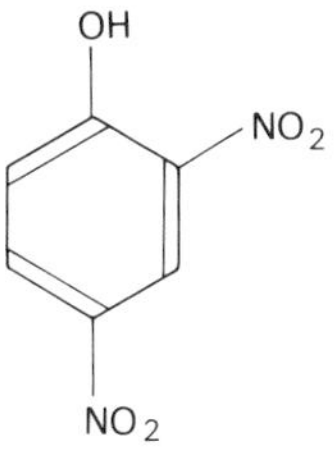

2,4-Dinitrophenol (DNP)

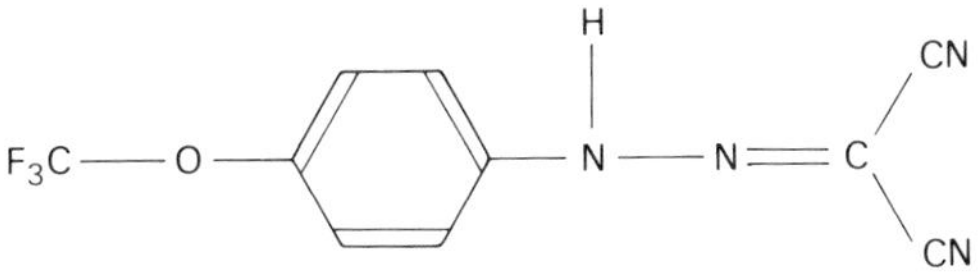

Carbonylcyanide *p*-trifluoromethoxyphenylhydrazone (FCCP)

Fig. 3.17 Structures of oxidative phosphorylation uncouplers 2,4-DNP and FCCP.

can be inhibited by oligomycin and dicyclohexylcarbodiimide (DCCD) (structure shown in Fig. 3.16). When ATP synthesis is coupled to electron flow the oligomycin or DCCD will inhibit both ATPase (which is blocked directly) and electron flow (because this is tightly coupled at ATP synthesis). There is another group which affects oxidative phosphorylation called uncouplers because they disconnect the coupling between ATP synthesis and electron transport. These are usually weak acids with an aromatic ring structure; the structures of two uncouplers 2,4-dinitrophenol (2,4-DNP) and *p*-trifluoromethoxycarbonylcyanide phenylhydrazone (FCCP) are shown in Fig. 3.17. These stimulate ATP hydrolysis in intact mitochondria and will relieve the inhibition by oligomycin on electron transport coupled to ATP synthesis, allowing electron transport to continue at an uncontrolled rate until the oxygen or substrate is exhausted (see Fig. 3.19(*a*) and (*b*)).

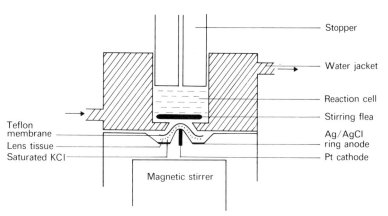

Fig. 3.18   The oxygen electrode.

3.6   Measurement of oxidative phosphorylation

Oxygen uptake by isolated mitochondria, cells or tissue slices, used generally to be monitored using Warburg manometers. While this apparatus gave
useful information concerning oxidative phosphorylation, initial rates of
oxygen consumption could not be measured, since only the total oxygen
uptake over a fixed period could be determined. In recent years the
oxygen electrode has been used to monitor oxygen consumption by
mitochondria and its use has led to a greater understanding of the mechanism of oxidative phosphorylation. A Clark-type oxygen electrode system is
shown in Fig. 3.18. The system has two electrodes, one platinum and one
silver mounted in the base of a cell surrounded by a water jacket.
Oxygen-saturated, isotonic, buffered medium is placed in the cell and
when the cap is put in place no further oxygen can dissolve into the
medium. The medium is separated from the electrodes by a thin Teflon
membrane which is permeable to oxygen; beneath this is a piece of paper
tissue soaked in saturated KCl which acts as an electrolyte. The platinum
electrode is polarised at -0.6 V with respect to the silver electrode and
oxygen is reduced at the platinum electrode as follows:

$$O_2 + 2H^+ + 2e^- \rightarrow H_2O_2$$
$$H_2O_2 + 2H^+ + 2e^- \rightarrow 2H_2O$$

Four electrons are consequently taken from the platinum electrode causing a current to flow from the Ag/AgCl electrode. The magnitude of the
current is directly proportional to the concentration (strictly the activity)
of the oxygen in solution. (The amount of oxygen actually reduced by the
platinum electrode is very small compared with that used by the mito-

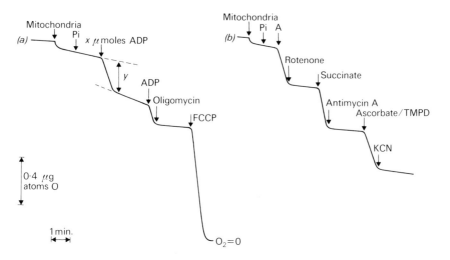

Fig. 3.19 Some typical results obtained with the oxygen electrode shown in Fig. 3.18.

The medium for both experiments contained KCl and Tris, pH 7.4 and if the mitochondria were prepared from rat liver, then the medium would also contain 5—10 mM malate and glutamate as substrate. At point A in (*b*) excess ADP (5*x* μmoles) or an uncoupler could be added.

chondria in the medium and is not usually taken into account.) Additions to the medium during the course of an experiment can be made with a syringe through the narrow hole in the cap and measurements can be made until the oxygen is used up.

Figure 3.19 shows two experimental traces obtained with a recording oxygen electrode. Figure 3.19(*a*) shows that there is a slow rate of oxygen consumption in coupled mitochondria until ADP is added, then oxygen is consumed rapidly until all of the ADP has been phosphorylated at which point the oxygen uptake rate slows down again. The rapid rate of oxygen uptake in the presence of ADP has been called State III by Chance and Williams and the state when ADP is exhausted is called State IV. The ratio of State III over State IV rates of oxygen uptake is called the respiratory control ratio (RCR) and this gives a measure of how well coupled the mitochondria are. A ratio of 1 would occur with uncoupled mitochondria; ratios of 3—6 are usual with malate or glutamate as substrates for rat liver mitochondria and ratios of 15 or so are not unusual with pyruvate as substrate for insect flight muscle mitochondria. Figure 3.19(*a*) also shows how oligomycin inhibits oxygen consumption but this can be relieved by the addition of the uncoupler FCCP, but then oxygen uptake is uncontrolled and only ceases when the system becomes anaerobic. Figure 3.19(*b*) demonstrates the action of the inhibitors rotenone, antimycin A and cyanide. Malate feeds in electrons at NADH and this is inhibited by rotenone. The inhibition can be overcome by the addition of succinate which feeds in electrons via the succinate dehydrogenase flavoprotein to ubiquinone. The succinate oxidation is inhibited by antimycin A, but this can be overcome by the addition of the artificial electron donor system ascorbate/tetramethyl-*p*-phenylenediamine (TMPD) which feeds in electrons to cytochrome oxidase which is, in turn, inhibited by cyanide. The basis for this experiment may be clarified by referring to Fig. 3.11.

3.7 The coupling of ATP synthesis to electron flow

Along the electron transport chain there are three coupling sites; this means that at each of these sites the passage of two electrons along the chain gives rise to the synthesis of one molecule of ATP. The three coupling sites are associated with Complexes I, III and IV respectively; there is no coupling site in Complex II (see Fig. 3.11). The number of coupling sites can be determined using an oxygen electrode. Figure 3.19(*a*) shows a single ADP induced stimulation of respiration followed by a State III to IV transition when ADP is exhausted. For every molecule of ADP added 1 ATP will be formed, thus if x μmoles of ADP are added we assume that when the State III → IV transition occurs x μmoles of ATP have been formed. If the amount of oxygen consumed in order to synthesise this ATP is y μg atoms (shown in Fig. 3.19(*a*)), the ratio of x/y is the ADP to 0 ratio; for a NADH-linked substrate (e.g. malate) this ratio should be approximately three and for a FAD-linked substrate (e.g. succinate) it will be two. This also means that for every pair of electrons (and protons) donated to the respiratory chain by malate 3 ATPs will be formed since one oxygen atom receives two electrons from cytochrome oxidase in order to form water. Consequently the ADP:O ratio is sometimes referred to as an ADP:$2e^-$ ratio. Similarly the ADP:$2e^-$ ratio for succinate oxidase activity (succinate oxidation by oxygen) is two. The most accurate determination is a P:O or P:$2e^-$ ratio in which the amount of phosphate used up when a certain amount of oxygen is consumed is assayed. The phosphate is assayed by complexing it with acidified ammonium molybdate followed by reduction to give a coloured complex which can be assayed by its absorption at 660 nm. It is usual to remove any ATP formed by including glucose and hexokinase which leads to glucose-6-phosphate formation.

Glucose + ATP → Glucose-6-phosphate + ADP

This is to prevent any ATP hydrolysis recycling phosphate. It also means that ADP is recycled and so State III respiration can be maintained for a long period. It is also possible to assay the glucose-6-phosphate formed, instead of the phosphate used up.

A knowledge of the free-energy changes associated with NADH oxidation by oxygen and ADP phosphorylation permit an estimation of the efficiency of the oxidative phosphorylation system in coupled mitochondria.

Exergonic respiration:
$$NADH + \tfrac{1}{2}O_2 + H^+ \rightarrow NAD^+ + H_2O \qquad G^{O'} = -220.3 \text{ kJ/mole}$$
Endergonic phosphorylation:
$$3ADP + 3Pi \rightarrow 3ATP + 3H_2O \qquad G^{O'} = + 91.5 \text{ kJ/mole}$$
Efficiency of energy transformation $= \dfrac{91.5}{220.3} \times 100 = 41.5$ per cent.

There are some mitochondria which are probably not always coupled, even *in vivo*. In brown adipose tissue mitochondria from hibernating animals the redox potential energy from the respiratory chain is not only used to produce ATP, but also to produce heat. This obviously has a very important role in temperature maintainance in these animals.

One of the fundamental questions is bioenergetics is how the energy released by electron transport from NADH to oxygen can be coupled to ATP synthesis from ADP and phosphate. Three theories have been put forward to try to explain this coupling. The first is the chemical coupling theory proposed by Slater in 1953; the chemiosmotic coupling theory was put forward in 1961 by Mitchell and the conformational theory in 1964 by Boyer.

3.7.1 The chemical coupling theory

This theory, which proposes a relatively simple sequence of exchange reactions for the coupling process, was based on the principles of substrate-level phosphorylation as illustrated by reaction sequence for glyceraldehyde-3-phosphate dehydrogenase in glycolysis (Fig. 3.20). In this reaction 1,3-diphosphoglycerate is formed which has a higher free energy of hydrolysis of the C-1 phosphate than that for the terminal phosphate of ATP (higher phosphate donor potential). It therefore reacts with ADP to form ATP a reaction catalysed by phosphoglycerate kinase. Slater consequently proposed that a chemical intermediate with high phosphate donor potential was formed in oxidative phosphorylation. In this theory a reduced electron carrier of the respiratory chain (e.g. AH_2) reacts with an oxidised carrier (e.g. B) adjacent to it (Fig. 3.21) and sufficient free-energy drop occurs to allow the reaction of A with an unknown compound C to give A-C. In sequential exchange reactions C is transferred to phosphate to give $C \sim P$, a phosphorylated intermediate and phosphate from $C \sim P$ to ADP to give ATP. During these reactions AH_2 is oxidised to A, B is reduced to BH_2 and C is recycled. A parallel scheme in

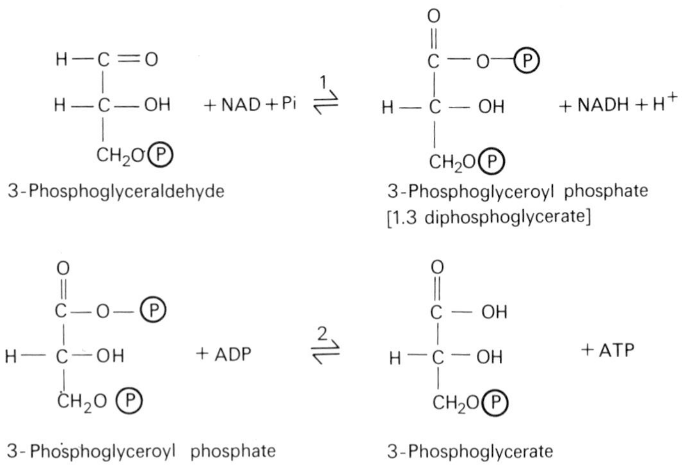

Fig. 3.20   The action of 3-phosphoglyceraldehyde dehydrogenase (1) and phosphoglycerate kinase (2).

which the initial 'high-energy' intermediate is formed with BH_2 instead of A is also shown in Fig. 3.21.

Confirmation of this theory would require the identification of $A \sim C$ (or $BH_2 \sim C$) and $C \sim P$. One of these intermediates ($C \sim P$) may be very unstable and difficult to observe, but it should be possible to observe spectroscopically any 'high-energy' intermediate involving the respiratory carriers. Since there are three coupling sites along the respiratory chain there would be three possible $A \sim C$ (or $BH_2 \sim C$) intermediates.

Any theory for the coupling of oxidative phosphorylation must explain the action of uncouplers and inhibitors. Uncouplers were originally thought to cause the breakdown of $A \sim C$ (or $BH_2 \sim C$) so that C is continually recycled permitting electron transport to occur without forming a phosphorylated intermediate. Oligomycin was thought to inhibit reaction (iii) (Fig. 3.21). The chemical theory promoted intensive research into the mechanism of oxidative phosphorylation, but no phosphorylated intermediates or 'high-energy' intermediates of the respiratory carriers have been incontravertibly demonstrated. It has been suggested that NAD^+ or ubiquinone might form phosphorylated derivatives involved in oxidative phosphorylation, but these claims have never been substantiated. Another inadequacy of the theory was that it did not explain why phosphorylation in mitochondria (and in chloroplasts and bacteria) occurs in membranous vesicles and why if these membranes are damaged the system becomes uncoupled.

72

(i) $AH_2 + B + C \rightleftharpoons A\sim C + BH_2$

(ii) $A\sim C + Pi \rightleftharpoons C\sim \text{\textcircled{P}} + A$

(iii) $C\sim \text{\textcircled{P}} + ADP \rightleftharpoons ATP + C$

A and B are electron carriers. An alternative scheme can be proposed in which a 'high-energy' intermediate $BH_2\sim C$ is formed instead of $A\sim C$:

(i) $AH_2 + B + C \rightleftharpoons A + BH_2\sim C$

(ii) $BH_2\sim C + Pi \rightleftharpoons C\sim \text{\textcircled{P}} + BH_2$

(iii) $C\sim \text{\textcircled{P}} + ADP \rightleftharpoons ATP + C$

Fig. 3.21   The chemical coupling theory.

Recently, potentiometric titrations (measurements of redox potentials of respiratory carriers in sites under different conditions) have indicated that cytochrome b-566, cytochrome a_3 and $[Fe\,S]_{1a}$, components associated with the three different coupling sites undergo a shift in midpoint potential when mitochondria are energised by adding either an oxidisable substrate or ATP. It has been suggested that these shifts in midpoint potential represent the formation of 'high-energy' intermediates. They could, however, equally well be explained as responses to changes in membrane potential or conformational changes in the membrane.

Griffiths and his co-workers have recently found evidence that lipoic acid may be involved in the ATPase complex. These workers consider that a 'high-energy' intermediate is formed between reduced lipoic acid and oleic acid (oleyl-lipoate) and that this subsequently donates an oleyl group to inorganic phosphate to give oleyl phosphate. This, in turn, may act as a phosphate donor to ADP. No linkage has been established between lipoic acid and the reactions of the electron transport scheme, but the fact that lipoic acid is itself an oxidation-reduction carrier suggests a possible linkage.

3.7.2 The chemiosmotic coupling theory

This theory, proposed by Mitchell in 1961 tried to overcome some of the drawbacks of the chemical theory and explain the need for membranous vesicles for oxidative phosphorylation. The electron transport chain was proposed to exist in 'loops', crossing the inner membrane from 'M' to 'C' sides and back again (Fig. 3.22). This arrangement agrees well with the experimentally determined organisation of the carriers (Fig. 3.13) having NADH and succinate dehydrogenases on the 'M' side, cytochrome c on the 'C' side and cytochrome oxidase spanning the membrane.

An important feature of the chain, according to the chemiosmotic hypothesis is that electron and hydrogen carriers alternate; e.g. the FMN of NADH dehydrogenase would carry hydrogens (protons + electrons) then an iron-sulphur protein would carry electrons, ubiquinone would carry hydrogens and cytochrome b electrons. Another crucial feature of

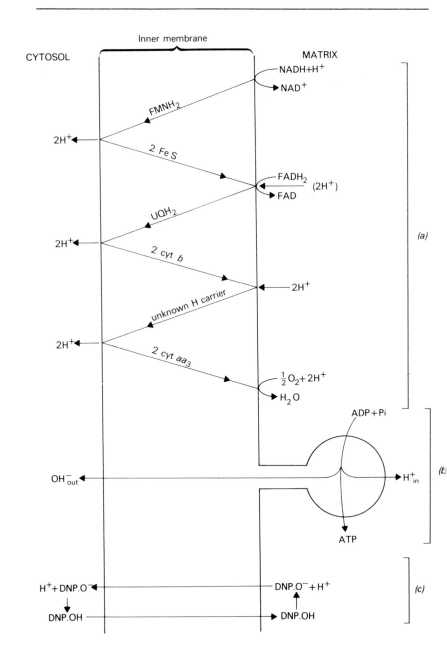

Fig. 3.22 The chemiosmotic hypothesis.
(*a*) Proton translocating electron transport chain.
(*b*) Proton translocating ATPase.
(*c*) Action of uncouplers.

the hypothesis is that the membrane must be impermeable to ions — especially H^+ and OH^-. The operation of the chain is considered to be as follows: hydrogens from NADH (formed by a TCA-cycle enzyme or other pyridine-nucleotide-linked enzyme) are transferred on the 'M' side to FMN, the prosthetic group from NADH dehydrogenase. On the 'C' side two hydrogen ions are released and two electrons go on down the respiratory chain; they are probably donated to iron-sulphur protein(s). At the 'M' side these electrons combine with two protons from the matrix, and are transferred to the 'C' side as reduced ubiquinone and so on until finally two electrons are donated to oxygen to form water. For each loop two hydrogen ions are released into the medium at the 'C' side. With succinate as the substrate only two loops are envisaged, $FADH_2$ feeding in directly to the second loop. So the operation of the respiratory chain would lead to proton translocation from the matrix to the cytosol and would establish a pH gradient across the mitochondrial membrane. Evidence that proton translocation occurred during respiration came from experiments carried out by Mitchell and Moyle in the 1960s. If mitochondria are suspended in buffered, isotonic medium in an anaerobic vessel and the pH is monitored, when a pulse of oxygen is added the medium rapidly acidifies. This acidification is equivalent to the translocation of $6H^+/O$ for an NAD^+-linked substrate or $4H^+/O$ for an FAD-linked substrate (Fig. 3.23). In fact the proton gradient established across the membrane has been called the protonmotive force (p.m.f.). This is reflected not solely as a pH gradient (chemical potential difference) as the experimentally determined p.m.f. is equivalent to 3.5 pH units which would mean that the matrix pH would have to be about 11. The p.m.f. is represented as having a chemical potential difference (Δ pH) component and an electrical potential difference ($\Delta\psi$) component where

p.m.f. $= \Delta\psi - z\Delta$ pH
($z = 2.303\ RT/F$ which approximates to 60 at 300 K)

The p.m.f. has been calculated to be about 230 mV in State IV respiration (mitochondria in 'energised' state).

Having established that the electron transport system can give rise to a transmembrane p.m.f., the problem remaining is how could this drive 'ATP' synthesis. During ATP synthesis from ADP and phosphate a molecule of water is also formed

ADP + Pi $\rightleftharpoons$ ATP + H_2O

The chemiosmotic hypothesis postulates that this water is formed aniso-

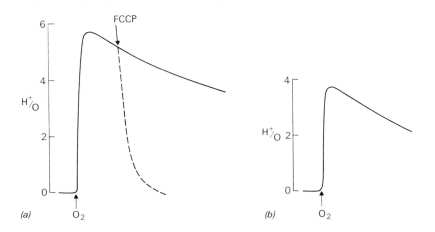

Fig. 3.23 The ejection of protons out of anaerobic mitochondria when oxygen is added. Mitochondria were incubated with the substrates (*a*) 3-hydroxybutyrate and (*b*) succinate.

tropically, that is an hydroxyl ion is released at the 'C' side of the inner membrane and a hydrogen ion at the 'M' side (Fig. 3.22).

$$ADP + Pi \rightleftharpoons ATP + H^+_{in} + OH^-_{out}$$

Thus the protonmotive force established by operation of the respiratory chain tends to draw protons and hydroxyls in specific directions from active site of ATP formation. This movement of a proton to the matrix and an hydroxyl to the cytosol would effectively neutralise the two protons ejected at each phosphorylation site during electron transport.

The coupling of electron transport to phosphorylation according to the chemiosmotic theory, depends on the integrity of the inner membrane as otherwise the protons translocated during electron transport would leak back and there would be no protonmotive force to drive ATP synthesis. Consequently it is considered that uncoupling agents, as they are usually lipid-soluble weak acids, would shuttle protons across the inner membrane (Fig. 3.22), causing a breakdown in the p.m.f. which would prevent ATP synthesis. Electron transport would be allowed to continue unchecked because the backpressure on electron transport, exerted by the p.m.f. in the absence of ADP or uncoupler, would be no longer operative. Oligomycin would act by blocking the anisotropic formation of H^+ and OH^-. Thus the p.m.f. could still be established by electron transport but could not be neutralised by ATP synthesis. Then the backpressure of the p.m.f. would prevent electron transport.

Mitchell has made a number of predictions based on the chemiosmotic

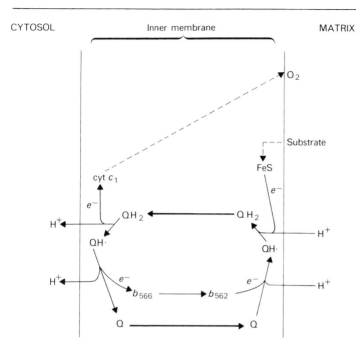

Fig. 3.24 The ubiquinone cycle (Q-cycle).

theory many of which have been shown to be correct; one important prediction was the existence of membrane transporters (see Ch. 4). If ions could diffuse across the membrane in an uncontrolled manner, the membrane potential would be broken down and no ATP would be synthesised, therefore the movement of ions must be controlled by specific membrane carriers.

Recent research has indicated that there may be three protons translocated at each electron transport loop rather than two. As only two protons per loop would be neutralised by the ATPase it is thought that the additional proton could be neutralised by the effect of the phosphate transporter (sect. 4.4) which will be operating during oxidative phosphorylation.

One problem associated with the chemiosmotic theory is that no hydrogen carrier was found for the third loop of the respiratory chain. Recently Mitchell has postulated the Q-cycle (ubiquinone cycle) (Fig. 3.24) to try to explain this without involving any unknown hydrogen carriers. In this cycle ubiquinone transfers protons in an electron transport driven process across the membrane. Reduced ubiquinone is oxidised in a two step process at the 'C' side of the membrane — there being a

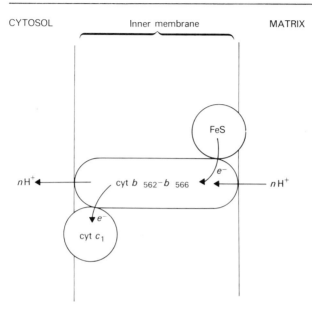

Fig. 3.25 Papa's 'vectorial Bohr' hypothesis for proton translocation during electron transport.

semiquinone intermediate (QH·). At each oxidation step a single proton is released to the intermembrane space and an electron is donated either to cytochrome c_1 (and then on down the terminal section of the chain) or to cytochrome b-566. The latter transfers the electrons back across the membrane via cytochrome b-562. This is then used to reduce Q → QH·. QH· is further reduced by QH_2 by an electron coming from the initial part of the electron transport chain. At both reduction steps a single proton is taken up from the matrix. The cycle is completed by the movement of oxidised and reduced ubiquinone in specific directions across the inner membrane. For each *single* electron passing through the 'Q'-cycle $2H^+$ are translocated giving a $H^+/2e^-$ ratio of four. Thus the Q-cycle would replace loops two and three. Besides giving a mechanism for proton transfer this scheme explains a number of difficulties encountered in understanding the kinetics of ubiquinone, cytochrome b and cytochrome c_1 oxidation-reduction.

A modified version of the chemiosmotic hypothesis (Fig. 3.25) has been proposed by Papa who suggests that energy-transforming electron carriers span the membrane (as is known to be the case for the cytochrome oxidase complex). Reduction of the prosthetic group of the carrier at 'M' side would be accompanied by protonation of the apoprotein. On reoxida-

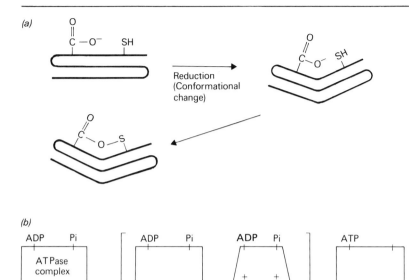

Fig. 3.26   Conformational coupling theory.
(*a*) Boyer's original formulation.
(*b*) Green's electromechanochemical model.

tion of the carrier at the 'C' side protons would be released from the apoprotein into the intermembrane space. This theory which Papa refers to as the vectorial Bohr theory has the merit that any stoichiometry of proton transfer to electron transfer can be accommodated.

3.7.3. The conformational coupling theory

The conformational coupling theory suggests that the energy released during electron transport processes is conserved in a conformational change in an electron carrier or complex of carriers. The idea was originally put forward by Boyer in 1964 who suggested that electron transport driven ATP synthesis might be analogous to a reversal of ATP driven contraction of actomysin in muscle.

Boyer suggested that perhaps a conformational change in an electron carrier brought a carboxyl and sulphydryl residue close enough to form an acyl-S linkage and that this was the 'high-energy intermediate' which could drive ATP synthesis (Fig. 3.26(*a*)).

Subsequent suggestions, such as Green's electromechanochemical coupling theory have postulated conformational changes in enzyme complexes rather than in individual carriers. Green suggests that a conformational change in one of the electron transport complexes, brought about by electron transport has mechanical and electrical strain components. These, in turn, induce similar mechanical and electrical changes in an ATPase complex and the reverse conformational change leads to ATP synthesis (Fig. 3.26(*b*)).

Hackenbrock and Green have observed electron microscopically, conformational changes in membrane and ATPase particles associated with changes in the energy state of the mitochondria. However, it cannot be ruled out that these are secondary results of osmotic changes resulting from energisation. In recent publications Green has gone to great lengths to try to accommodate the electromechanochemical theory to most of the known properties of the oxidative phosphorylation system. However, a major problem is that the kinetics of conformational change, which have been studied by Radda using the fluorescent membrane probe anilinonaphthalene sulphonic acid (ANS) seem to be too slow for the observed changes in conformation to be the primary energy conserving event.

3.7.4 Discussion of the coupling mechanism

All three major theories for oxidative phosphorylation agree that coupling involves firstly formation of some kind of energised state and that this drives ATP synthesis.

	Energised state	
Chemical	Electron transport → Covalent intermediate	→ ATP
Chemiosmotic	Electron transport → Protonmotive force	→ ATP
Conformational	Electron transport → Conformational change	→ ATP

It is the nature of the energised state which is at present in dispute.

It is the view of the authors that, at present, the chemiosmotic hypothesis offers the most acceptable explanation of the available data. However, even if this view is correct, it is clear that much more work must be done to work out the details of the process.

One of the most convincing pieces of evidence in favour of the chemiosmotic hypothesis stems from the work of Stoeckenius on the purple membrane of *Halobacterium halobium*. When this bacterium is grown under conditions of low aeration and high illumination patches of membrane containing the purple protein bacteriorhodopsin appear (purple membrane). Stoeckenius has shown that illumination of the bacteria leads

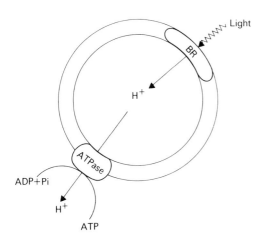

Fig. 3.27 ATP synthesis in phospholipid vesicles with incorporated *H. halobium* purple membrane and mitochondrial ATPase.
BR = bacteriorhodopsin.

to bacteriorhodopsin driving a proton extrusion from inside the bacteria to outside and that this leads to ATP synthesis. Furthermore, Racker and Stoeckenius have prepared phospholipid vesicles and have incorporated into these purple membrane and mitochondrial F_1-ATPase + OSCP (see sect. 3.8). On illumination the purple membrane causes proton uptake (because its orientation is opposite to that in the bacterium) and the proton gradient so formed drives ATP synthesis by the mitochondrial ATPase (Fig.3.27).

3.8   The adenosine triphosphatase complex

As briefly discussed in Chapter 1 (Fig. 1.8) in 1962 Fernandez-Moran showed by observing negatively stained mitochondria under the electron microscope that there were spheres attached to the inside (M) side of the inner mitochondrial membrane. It is now generally accepted that these spheres are the adenosine triphosphatase complex. Evidence for this came from the work of Racker who prepared submitochondrial particles by shaking mitochondria with glass beads. When the intact mitochondria were removed from the sample by centrifugation the supernatant had oxidative phosphorylation activity. The supernatant was then centrifuged at approximately 100 000 g and the pellet contained submitochondrial particles, which could oxidise substrates, but were unable to phosphorylate ADP, they had also lost a lot of the membrane spheres. If the supernatant was added back to the submitochondrial particles oxidative phosphorylation was restored. Many proteins have been isolated from this supernatant and they have been called coupling factors, because they restore the coupling between oxidation and phosphorylation. The first coupling factor isolated and studied in detail was called F_1 and the use of anti-F_1 antibodies proved that F_1 contained ATPase activity. This protein is spherical and is generally associated with the spheres seen attached to the inner mitochondrial membrane. It is usually referred to as the F_1-ATPase. In fact this protein does not have the characteristics of the ATPase *in vivo*.

Another coupling factor which was isolated called F_0 (also called oligomycin-sensitivity conferring protein, OSCP) is accepted as representing the stalk of the ATPase seen in the electron microscope and as its name implies it confers oligomycin sensitivity on the F_1-ATPase. In fact it probably does this by binding the F_1 to the inner membrane. The F_1-ATPase plus F_0 plus a section of the membrane is called the tripartite unit or oligomycin-sensitive ATPase and this has a large number of the characteristics of the ATPase *in vivo*.

The reaction catalysed by this enzyme complex *in vivo* is

$$ADP + Pi \rightleftharpoons ATP + H^+_{in} + OH^-_{out}$$

('in' and 'out' refer to the M and C sides of the membrane respectively). *In vivo* and in isolated (coupled) mitochondria ATP synthesis is the favoured reaction, while in the presence of uncouplers and in an isolated purified ATPase preparation ATP hydrolysis is the favoured reaction. Thus ATP synthesis occurs when the membrane is intact, since when the membrane is impermeable to hydrogen ions the water formed in the ATPase reaction is formed anisotropically because of the protonmotive force (p.m.f.) formed during electron transport (sect. 3.7.2). The hydroxyl ions are moved out of the matrix pulling the reaction in the direction of ATP synthesis. This is in contrast to other ATPases, e.g. Ca^{++}-dependent ATPase in muscle and $Na^+ - K^+$-dependent ATPase in plasma membrane, which catalyse ATP hydrolysis *in vivo*. The mitochondrial ATPase is therefore sometimes referred to as ATP synthetase.

The enzyme can be assayed discontinuously by reacting the ATP with luciferin. This reaction is catalysed by the enzyme luciferase as follows:

Luciferin + ATP $\rightarrow$ Adenyl-luciferin + PPi (pyrophosphate)
Adenyl-luciferin + O_2 $\rightarrow$ Adenyl-oxyluciferin + H_2O + light

The light emitted is proportional to the amount of ATP present and can be measured using one photomultiplier in a scintillation counter. Synthesis of ATP can also be assayed discontinuously by following the incorporation of [^{32}P] orthophosphate into ATP. The inorganic and organic (ATP) phosphates can be separated by adding acidified molybdate, which complexes the inorganic phosphate, this can then be extracted into a mixture of isobutanol and benzene leaving the ATP (and ADP) in the aqueous phase.

The hydrolysis of ATP can be assayed discontinuously by monitoring

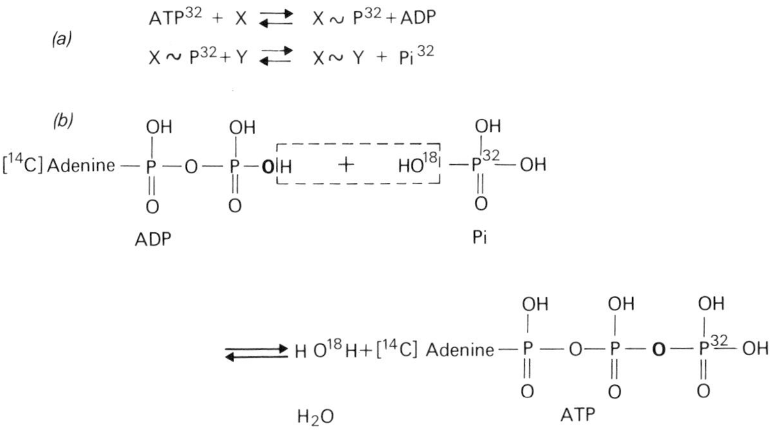

(a)

$$ATP^{32} + X \rightleftharpoons X \sim P^{32} + ADP$$

$$X \sim P^{32} + Y \rightleftharpoons X \sim Y + Pi^{32}$$

(b)

[^{14}C] Adenine — P(OH)(=O) — O — P(OH)(=O) — O|H + HO^{18}| — P^{32}(OH)(=O) — OH

ADP + Pi

$\rightleftharpoons$ H O^{18} H + [^{14}C] Adenine — P(OH)(=O) — O — P(OH)(=O) — O — P^{32}(OH)(=O) — OH

H_2O + ATP

Fig. 3.28 The exchange reactions.

the appearance of phosphate, this is complexed with molybdate as previously described (sect. 3.7). The hydrolysis of ATP can also be measured continuously by measuring the extrusion of hydrogen ions using a pH electrode. The ATPase complex in intact mitochondria is inhibited by oligomycin, rutamycin and dicyclohexylcarbodiimide (DCCD).

3.8.1 The exchange reactions

An exchange reaction is one where radiolabel can be incorporated into one compound from another without overall metabolism of either compound. The first exchange reaction to be examined in detail in mitochondria was the phosphate-ATP exchange (Pi-ATP). Mitochondria were incubated, in absence of substrate, with ^{32}Pi, ATP and Mg^{++} and the ATP became labelled with [^{32}P] and no oxygen was consumed. The mechanism proposed for such an exchange is shown in Fig. 3.28(a), X and Y could be protein components of the ATPase complex, that is a substrate/enzyme complex is formed. This exchange is inhibited by uncouplers. This and other exchange reactions are illustrated in Fig. 3.28(b). When ADP, labelled with [^{14}C] in the adenine ring, is added to mitochondria ATP becomes labelled with [^{14}C] and if water is labelled with [^{18}O] then inorganic phosphate becomes labelled by [^{18}O]. These are called the ADP-ATP and Pi-H$_2$O exchange reactions. It has also been shown that the bridge oxygen in ATP comes from ADP and not inorganic phosphate.

These exchange reactions are considered to be partial reactions of oxidative phosphorylation and they illustrate the reversibility of oxidative phosphorylation (see also sect. 3.9). These reactions only operate if the

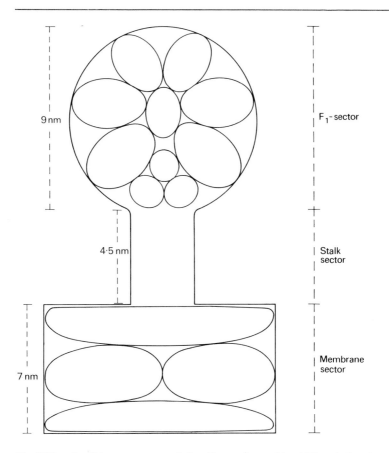

Fig. 3.29   Possible arrangement of the oligomycin-sensitive ATPase (tripartite repeating unit). (Adapted from Fig. 2 of Senior, A. E. *Biochem. Biophys. Acta*, **301** (1973) 249.)

'coupling device' (energised state) is present since they are inhibited by uncouplers.

Between 1955 and 1965, exchange reactions were studied intensively to try and gain information about oxidative phosphorylation. However, now the study of the complete oxidative phosphorylation system is favoured.

3.8.2 The isolated oligomycin-sensitive ATPase complex
As previously mentioned the oligomycin-sensitive ATPase consists of three parts, the F_1-ATPase, F_0 (OSCP) and a membrane section. Most of the work on the isolated complex has been carried out using a preparation from beef heart mitochondria, but the yeast mitochondrial preparation

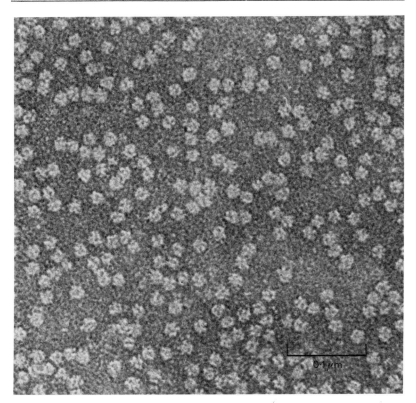

Fig. 3.30   Electron micrograph of F$_1$ isolated from beef heart mitochondria. The preparation was negatively stained with phosphotungstate. (Courtesy of Prof E. Racker.)

shows similar characteristics. The preparation involves treatment of sub-mitochondrial particles with deoxycholate, which yields a fraction containing ATPase activity plus some cytochromes and flavoproteins. These are precipitated out by dialysis and the ATPase is precipitated on a sephadex column and further purified by cholate and ammonium sulphate fractionation. The preparation contains lipid, but phospholipid has to be added back to some preparations for full activity. Its approximate molecular weight is 468 000.

   This purified complex has many of the properties of the ATPase in submitochondrial particles. It has a pH optimum of 8–9, requires Mg^{++} or Mn^{++} (but not Ca^{++}) for activity and is not cold labile (this is a characteristic of the isolated F$_1$-ATPase). Its nucleotide specificity is ATP > GTP, ITP > UTP, CTP. ATP hydrolysis catalysed by this complex is inhibited by

Table 3.6 The components of the ATPase complex
(The molecular weights given for F_1-ATPase are those of the beef heart mitochondrial enzyme. The values for OSCP and the membrane sector proteins are those of the yeast mitochondrial complex.)

	Protein	Molecular weight	Molecular weight contribution to complex
F_1	$F_1.1$	53 000	159 000
	$F_1.2$	50 000	150 000
	$F_1.3$	33 000	33 000
	$F_1.4$	17 000	17 000
	$F_1.5$	7 500	7 500
	Mitochondrial inhibitor (MI)	10 000	10 000
	Molecular weight of F_1 = 376 500		
F_0	OSCP	18 000	18 000
Membrane sector	MS.1	29 000	29 000
	MS.2	22 000	22 000
	MS.3	12 000	12 000
	MS.4	7 800	7 800
	Total molecular weight of complex = 464 300		

oligomycin, rutamycin, DCCD, aurovertin, mitochondrial inhibitor (discussed in sect. 3.8.3), tributyl tin chloride, mercurials and ADP, all at concentrations which inhibit oxidative phosphorylation in mitochondria and submitochondrial particles. Two exchange reactions (ATP-Pi and ATP-ADP) are not catalysed by this preparation.

3.8.3 The structure of the ATPase complex

The structure of the tripartite unit is shown in Fig. 3.29. The F_1-ATPase is spherical with a diameter of c. 9 nm, F_0 (OSCP) is about 4.5 nm long and the membrane sector is about 7 nm wide. All three parts of the oligomycin-sensitive ATPase have been studied in detail to determine their protein composition.

F_1-ATPase has ATPase (ATP hydrolysis) activity which is inhibited by aurovertin, mitochondrial inhibitor and ADP, but not oligomycin or DCCD. It is prepared by long sonication of mitochondria or submitochondrial particles and purified F_1-ATPase does not require phospholipids for activity. It can use Ca^{++} as an activator and is as active with GTP and ITP as it is with ATP. F_1-ATPase as a molecular weight of c. 360 000, it is soluble in water and cold labile. Its inactivation at low temperature means that preparations of F_1 have to be carried out at room temperature. An electron micrograph of purified F_1-ATPase is shown in Fig. 3.30. This gives an indication of the subunit structure of this complex.

When F_1 is analysed by SDS gel electrophoresis 5–6 protein bands can be detected after coomassie blue or amido black staining. Five of these bands have been called respectively subunits 1, 2, 3, 4 and 5 their molecular weights in the beef heart mitochondrial F_1 are shown in Table 3.6, but these subunits have similar molecular weights when isolated from beef heart, yeast and rat liver mitochondria (and chloroplasts). The sixth band is the mitochondrial inhibitor (MI) this protein inhibits ATP hydrolysis, but not synthesis in mitochondrial preparations. When this protein binds to F_1, which it does readily in the presence of ATP and MG^{++}, the F_1-ATPase is no longer cold labile. F_1 is thought to contain three subunits of $F_1.1$, three subunits of $F_1.2$ and one subunit each of $F_1.3$, $F_1.4$, $F_1.5$ and MI, giving a molecular weight of 376 500, very near to that of the isolated complex.

F_0 or oligomycin sensitivity conferring protein has been isolated as a cylindrical protein of *c.* 4.5 nm length and *c.* 3 nm diameter with a molecular weight of 18 000. It has no known enzymic function, but is required to bind F_1 back onto submitochondrial particle membranes.

The membrane sector of the tripartite unit is the most difficult to purify since the proteins are very hydrophobic and crude preparations contain a large amount of phospholipid. It has been postulated that there are a minimum of four proteins present in this sector. If there was only one of each of these proteins in the tripartite unit the total molecular weight would be 464 300 (Table 3.6) which is close to the measured molecular weight of 468 000.

A possible structural arrangement of the protein subunits of the oligomycin-sensitive ATPase is shown in Fig. 3.29. It has been suggested that the large number of subunits in the tripartite unit are necessary so that the ATPase can undergo large conformational changes which may be necessary for its activity. The inhibitor aurovertin will be useful for the study of the mechanism of ATPase, since it is a potent inhibitor of oxidative phosphorylation, interacting directly with F_1-ATPase forming a fluorescent complex.

3.9   Reversed electron flow

The electron transport driven synthesis of ATP is a fully reversible process. It has been shown that the hydrolysis of ATP to ADP and Pi will drive the electron transport system in reverse against the redox potential gradient. Reversal of practically all sections of the electron transport chain can be achieved but the best studied system involves the reduction of NAD^+ to NADH by succinate. The pathway of electron transport from succinate to NAD^+ is indicated in Fig. 3.31. The normal direction of electron transport

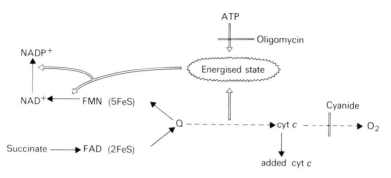

Fig. 3.31   Reversed electron transport and energy-linked transhydrogenase.

is operating from succinate to ubiquinone (Complex II) but the pathway from ubiquinone to NAD^+ (Complex I) is operating in the opposite direction from its usual mode. Reversal of the phosphorylation step in Complex I requires a considerable energy input. This can either be provided by ATP hydrolysis or by operation of the phosphorylation step in Complex III. When electron transport through complex III is used as an energy source oligomycin does not affect reversal of electron flow through Complex I. Therefore ATP synthesis is not a necessary prerequisite to reversed electron flow and the mitochondrial energised state must mediate the process. ATP driven reversal is, on the other hand, sensitive to oligomycin. As uncouplers cause the dissipation of the energised state these inhibit reversed electron transport driven by either ATP hydrolysis or electron transport through Complex III.

3.10 NAD(P)⁺ transhydrogenase

The enzyme $NAD(P)^+$ transhydrogenase (pyridine nucleotide transhydrogenase) (EC.1.6.1.1) catalyses the reaction $NADH + NADP^+ \rightleftharpoons NAD^+ + NADPH$ and is usually associated with the mitochondrial inner membrane. The reaction can be driven in the direction of NADPH formation by coupling it to the hydrolysis of ATP (Fig. 3.31). It is also similar to reversed electron transport in that conventional electron transport will drive NADPH formation by an oligomycin-insensitive pathway. Mitchell has postulated that the transhydrogenase may comprise a fourth proton translocating loop (Loop 0) of the respiratory chain. In adrenal mitochondria it is thought that reversed electron transport and $NAD(P)^+$ transhydrogenase may provide the means of transferring reducing potential from the TCA cycle to NADPH which may be used in steroid hydroxylation by the P450 system.

Suggested further reading

Books

JONES, C. W. (1976) *Biological Energy Conservation. Outline Studies in Biology Series.* Chapman & Hall.

LEMBERG, R. and BARRETT, J. (1973) *Cytochromes.* Academic Press, New York and London.

LLOYD, D. (1974) *The Mitochondria of Microorganisms.* Academic Press, London and New York.

NICHOLLS, P. (1975) *Cytochromes and Biological Oxidation.* Oxford Biology *Reader* No. 66.

RACKER, E. (1965) *Mechanisms in Bioenergetics.* Academic Press, New York and London.

RACKER, E. (1974) *A New Look at Mechanisms in Bioenergetics.* Academic Press, New York and London.

The coupling theories

BOYER, P. D. (1965) Carboxy activation as a possible common reaction in substrate-level and oxidative phosphorylation and muscle contraction, in *Oxidases and Related Redox Systems,* 2, 994–1004, Wiley.

GREEN, D. E. and JI, S. (1972) The electromechanochemical model of mitochondrial structure and function, *Bioenergetics,* 3, 159–202.

GRIFFITHS, D. E. and HYAMS, R. L. (1977) Oxidative phosphorylation: A role of lipoic acid and unsaturated fatty acids, *Biochem. Soc. Trans.,* 5, 207–8.

HACKENBROCK, C. R. (1966) Ultrastructural bases for metabolically linked mechanical activity in mitochondria, *J. Cell Biol.,* 30, 269–97.

HATEFI, Y. and HANSTEIN, W. G. (1972) On energy conservation and transfer in mitochondria, *Bioenergetics,* 3, 129–36.

MITCHELL, P. (1966) Chemiosmotic coupling in oxidative and photosynthetic phosphorylation, *Biol. Rev.,* 41, 445–502.

MITCHELL, P. (1976) Possible molecular mechanisms of the protonmotive function in cytochrome systems, *J. Theor. Biol.,* 62, 327–67.

PAPA, S. (1976) Proton translocation reactions in the respiratory chains, *Biochim. Biophys. Acta,* 456, 39–84.

SLATER, E. C. (1953) Mechanism of phosphorylation in the respiratory chain, *Nature,* 172, 975–8.

Reviews on ATPase

BEECHEY, R. B. (1974) Structural aspects of mitochondrial ATPase, *Biochem. Soc. Trans.,* 2, 466–71.

PANET, R. and SANADI, D. R. (1976) Soluble and membrane ATPase of mitochondria, chloroplasts and bacteria: Molecular structure, enzymic properties and functions, *Curr. Topics in Mem. and Trans.,* 8, 99–160.

RACKER, E. (1970) The two faces of the inner mitochondrial membrane system, *Essays in Biochem.,* 6, 1–22.

SENIOR, A. E. (1973) The structure of mitochondrial ATPase, *Biochim. Biophys. Acta,* 301, 249–77.

TZAGOLOFF, A. (1971) Structure and biosynthesis of the membrane ATPase of mitochondria, *Curr. Topics in Mem. and Trans.,* 2, 157–206.

Other topics

BALTSCHEFFSKY, H. and BALTSCHEFFSKY, M. (1974) Electron transport phosphorylation, *Ann. Rev. Biochem.,* 43, 871–97.

BOXER, D. H. (1975) The location of the major polypeptide of the ox heart mitochondrial inner membrane, *FEBS Letts.*, **59**, 149—51.

CHANCE, B. and WILLIAMS, G. R. (1956) The respiratory chain and oxidative phosphorylation, *Adv. Enzymol.*, **17**, 65—134.

CHAPPELL, J. B. (1961) Integrated oxidations in isolated mitochondria, *Proc. 1st. IUB/IUBS Int. Sym.*, **11**, 71—83.

DAWSON, A. and SELWYN, M. (1974) in *Companion to Biochemistry*, pp. 558—86, eds. J. R. Lagnada *et al.*

DEAMER, D. W. (1969) ATP synthesis: The current controversy, *J. Chem. Ed.*, **46**, 198—206.

GREVILLE, G. D. (1969) A scrutiny of Mitchell's chemiosmotic hypothesis, *Curr. Topics Bioenerg.*, **3**, 1—78.

NICHOLLS, D. G. (1976) Bioenergetics of brown adipose tissue mitochondria, *FEBS Letts.*, **61**, 103—10.

RACKER, E. and STOECKENIUS, W. (1974) Reconstitution of purple membrane vesicles catalysing light-driven proton uptake and ATP formation, *J. Biol. Chem.*, **249**, 662—3.

SKULACHEV, V. P. (1971) Energy transformations in the respiratory chain, *Curr. Topics Bioenerg.*, **4**, 127—90.

VAN DAM, K. and MEYER, A. J. (1971) Oxidation and energy conservation by mitochondria, *Ann. Rev. Biochem.*, **40**, 115—60.

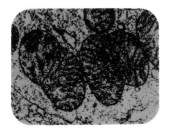

Chapter 4

Transport of ions across
the mitochondrial membrane

4.1 Introduction

The structure of the mitochondrion has been discussed in Chapter 1. It is
generally accepted that the mitochondrial outer membrane is freely permeable to small molecules, e.g. sucrose, adenine nucleotides and coenzyme
A, but is impermeable to larger molecules, e.g. large carbohydrates such as
inulin and proteins such as albumin. The permeability properties of the
inner mitochondrial membrane are not so straightforward. Since all membranes contain phospholipid (20–40 per cent of the mitochondrial inner
membrane) we can gain some information by comparing the permeability
properties of the inner membrane with that of articificial phospholipid
membranes. Artificial phospholipid membranes are permeable to chloride
ions, but less permeable to nitrate, sulphate and phosphate ions. This
would imply that membranes were only permeable to anions of a certain
hydrated size, 2.9 Å, e.g. Cl⁻. The mitochondrial inner membrane is unusual because it contains a high percentage of cardiolipin (10–20 per cent
of the total phospholipid). If cardiolipin is incorporated into artificial
phospholipid membranes they lose their permeability to Cl⁻ ions. The
inner mitochondrial membrane is permeable to large anions, e.g. phosphate
and acetate, but is not permeable to chloride ions. The chloride ion
impermeability could be explained by the presence of cardiolipin, but the
permeability of phosphate was explained by postulating the existence of a
specific carrier in the membrane.

The site of action of some enzymes of various metabolic pathways
within the cell also necessitates the transfer of some other fairly large
charged molecules across the inner mitochondrial membrane. The enzymes

for the conversion of glucose to pyruvate (glycolysis) are in the cytosol, but those for the complete oxidation of pyruvate to CO_2 and H_2O (TCA cycle, Ch. 2) are in the mitochondrial matrix. The enzymes for fatty acid biosynthesis (the fatty acid synthetase complex) are in the cytosol, but those for fatty acid oxidation are all in the matrix (Ch. 2). Some enzymes of the urea cycle are in the matrix and some are in the cytosol (Ch. 2) and ATP is formed in the matrix by the action of ATPase (Ch. 3), but is required for metabolic functions mainly in the rest of the cell.

These reactions indicate that at least the molecules pyruvate, fatty acyl CoA, acetyl CoA, citrulline, ornithine, ATP and ADP must cross the inner mitochondrial membrane. This is achieved by specific membrane transporters or carriers which are probably specialised proteins, and these allow the passage of large charged molecules across the inner mitochondrial membrane.

4.2 Methods of determining mitochondrial permeability

The two main methods for determining whether mitochondria are permeable to anions are the 'spaces' technique and the 'ammonium swelling' technique. The 'spaces' technique compares the amount of penetration of the radiolabelled anion into mitochondria with that of tritiated water and ^{14}C-labelled sucrose. T_2O can penetrate into the whole mitochondria and ^{14}C-labelled sucrose can only penetrate up to, but not into, the matrix. The metabolism of the anion under investigation has to be inhibited in order to investigate only its permeability properties. This can often be achieved using rotenone and/or antimycin A (Ch. 3). The mitochondria are incubated in media containing radiolabel and the reaction is stopped by centrifugation. This is sometimes achieved by centrifuging the mitochondria through a layer of silicone oil. From the radioactivity in the pellet fraction it is possible to determine whether the anion under investigation has penetrated into the matrix or not. To determine very detailed kinetics for a transporter the 'inhibitor-stop' technique is used, then the reaction is stopped using an inhibitor of the transporter, this allows measurements to be made at very short time intervals. The mitochondria can then be separated from the supernatant by centrifugation later. One advantage of the 'spaces' technique is that only very small quantities of material need be used.

The 'ammonium swelling' technique is a rapid method for obtaining qualitative information about transporters. Mitochondria are placed in an iso-osmotic ammonium solution of the anion under investigation and the optical density at 640 nm is monitored continuously. NH_3 crosses the membrane (Fig. 4.1) by diffusion down a concentration gradient. Then the NH_3 associates with protons leaving an excess of hydroxyl ions in the

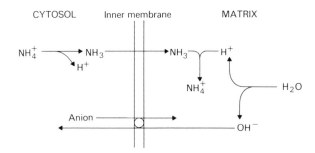

Fig. 4.1   The ammonium swelling technique.

matrix which could exchange with a permeant anion. There is an increase in the number of ions, and therefore the osmotic pressure, in the matrix, since H_2O has been replaced with OH^- or an anion and NH_4^+, therefore the mitochondria swell due to osmosis. A decrease in light scattering, i.e. a decrease in optical density, at 640 nm results when the mitochondria swell due to the penetration of ammonia and the anion. The swelling will only be observed by this method if the anion exchanges for hydroxyl ions, e.g. phosphate, otherwise there will be no increase in the number of ions in the matrix, since hydroxyl ions will not be removed from the matrix and dissociation of water will not take place. An anion/anion exchange can only be observed by this method if it is coupled to an exchange with hydroxyl ions, e.g. dicarboxylate and tricarboxylate transporters (see Fig. 4.6), since a one for one exchange would not require NH_3 to cross the membrane and not cause an increase in the number of ions in the matrix. The ammonium salts can be replaced by K^+ salts but since K^+ cannot cross the membrane valinomycin has to be added to facilitate K^+ entry (see later). Figure 4.2 shows some typical results obtained with this method. In Fig. 4.2(*a*) when mitochondria are added to an iso-osmotic solution of ammonium phosphate they swell rapidly, because the anion and cation can both penetrate into the matrix. In Fig. 4.2(*b*) the mitochondria do not swell in potassium phosphate until valinomycin and uncoupler are added to allow potassium across the membrane. Mitochondria do not swell in iso-osmotic NH_4Cl because chloride is impermeant, and they will not swell in iso-osmotic ammonium malate until a catalytic amount of phosphate is added. In all these experiments mitochondria were inhibited with rotenone to prevent metabolism of the anions under investigation.

Another method is to monitor the reduction of NAD^+ by the anion. This method is useful for anions whose metabolism is NAD^+-linked, e.g. isocitrate and glutamate (Fig. 4.3(*a*) and (*b*)). Figure 4.4 shows a typical experiment where mitochondria are placed in medium in a fluorimeter.

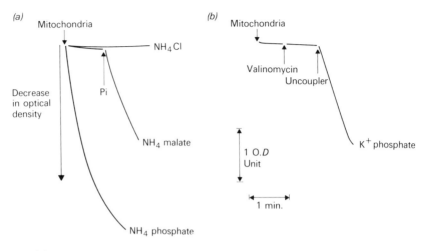

Fig. 4.2 Some results obtained using the swelling technique.

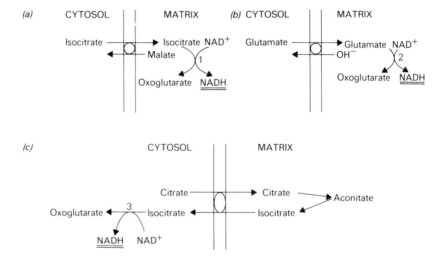

Fig. 4.3 The reduction of NAD^+ as a measure of transport.
1. Isocitrate dehydrogenase. 2. Glutamate dehydrogenase. 3. Added NAD^+-linked isocitrate dehydrogenase.

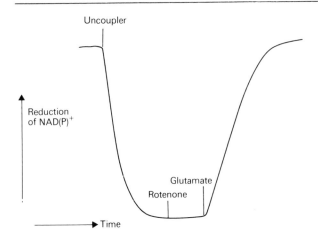

Fig. 4.4   The reduction of intramitochondrial NAD$^+$ by glutamate monitored by fluorescence.

The matrix NADH is fully oxidised by the addition of uncoupler then rotenone (or antimycin A) is added to prevent further metabolism of NADH via the electron transport chain. Glutamate (or isocitrate) is then added and if it penetrates into the matrix reduction of NAD$^+$ will be monitored in the fluorimeter at 480 nm. The tricarboxylate/tricarboxylate exchange has been observed by adding citrate to whole mitochondria then monitoring the reduction of external NAD$^+$ at 340 nm, in the presence of added isocitrate dehydrogenase, as isocitrate is transported out of the matrix (see Fig. 4.3(c)).

4.3   Types of transporter

There are two main types of transporter in mitochondria: uniporters (Fig. 4.5(a)), where one ion is transported into the matrix and no other ions are associated with this movement and antiporters (Fig. 4.5(b)) where one ion is moved into the matrix in exchange for another of similar charge.

As described in Chapter 3 a protonmotive force is set up across the mitochondrial membrane during electron transport, this gives rise to the formation of a negative charge on the matrix side of the membrane. The p.m.f. can be used to exchange ions of unlike charge across the membrane and to accumulate ions against a concentration gradient. This transport is called electrogenic transport. Figures 4.5(c) and 4.5(d) show an electrogenic uniporter and two possible electrogenic antiporters. An electrogenic transporter can only operate in one direction, since its reversal would not be favoured by the membrane potential. The membrane potential is

Table 4.1 Mitochondrial transporting systems

Transporter	Probable *in vivo* function (Out → In)	Inhibitors	Probable biological importance
Phosphate	Phosphate (out) / OH^- (in)	*N.* ethyl maleimide, mersalyl	1. Mitochondrial ATP synthesis 2. Allowing flux of dicarboxylates and therefore tricarboxylates
Adenine nucleotide	ADP^{3-} (out) / ATP^{4-} (in)	Atractyloside, CO-atractyloside, Bongkrekic acid	In mitochondrial ATP synthesis
Pyruvate	Pyruvate (out) / OH^- (in)	α-Cyano (4.OH) cinnamate	The link between glycolysis and the TCA cycle
Dicarboxylate (Malate and Succinate)	Malate (out) / Phosphate **or** Succinate (in); Citrate (out) / Malate (in)	n-Butyl malonate, 2-Phenyl succinate	1. Transferring reducing equivalents into and out of the matrix 2. Provides C-skeleton for PEP
Tricarboxylate (Citrate and isocitrate)	Malate or Isomalate **or** Citrate (out) / Isocitrate (in)	Citrate analogues	1. Transfer of acetyl CoA into cytosol 2. Control of PFK 3. Provision of NADPH in cytosol

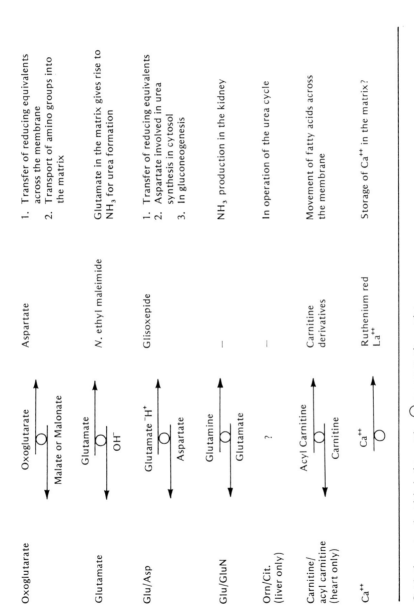

Transporter	Scheme	Inhibitor	Function
Oxoglutarate	Oxoglutarate ⇄ Malate or Malonate	Aspartate	1. Transfer of reducing equivalents across the membrane 2. Transport of amino groups into the matrix
Glutamate	Glutamate ⇄ OH^-	*N*. ethyl maleimide	Glutamate in the matrix gives rise to NH_3 for urea formation
Glu/Asp	Glutamate $^-H^+$ ⇄ Aspartate	Glisoxepide	1. Transfer of reducing equivalents 2. Aspartate involved in urea synthesis in cytosol 3. In gluconeogenesis
Glu/GluN	Glutamine ⇄ Glutamate	—	NH_3 production in the kidney
Orn/Cit. (liver only)	?	—	In operation of the urea cycle
Carnitine/ acyl carnitine (heart only)	Acyl Carnitine ⇄ Carnitine	Carnitine derivatives	Movement of fatty acids across the membrane
Ca^{++}	Ca^{++}	Ruthenium red La^{++}	Storage of Ca^{++} in the matrix?

Out is the cytosol and *In* is the matrix. ○ represents the carrier.

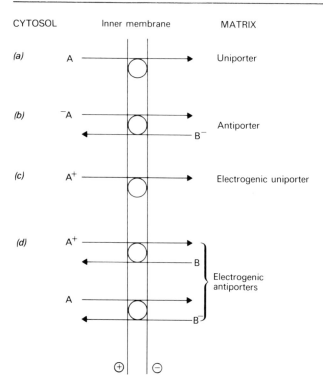

Fig. 4.5 Different types of transporter.

probably used to transport ions in a biologically favourable direction, e.g. adenine nucleotide exchange, glutamine/glutamate exchange (see sect. 4.7).

4.4 Anion transporters

The transporters so far characterised for the rat liver mitochondrial inner membrane are listed in Table 4.1 with their inhibitors, probable mode of action and probable biological importance. Also listed are some anion transporters found in other tissues and the calcium transporter. Much of the early work using the swelling technique was done by Chappell and co-workers, although a great deal is known about the adenine nucleotide transporter from the work of Klingenberg, Vignais and co-workers.

The phosphate and adenine nucleotide transporters probably occur in mitochondria from all tissues because of their involvement in ATP synthesis, the main function of mitochondria. *Phosphate* is transported into the matrix in exchange for hydroxyl ions, and this exchange is inhibited by several sulphydryl reagents, e.g. *N*-ethyl maleimide (NEM) and mersalyl. The phosphate transporter is capable of causing an accumulation of phosphate against a concentration gradient, in blowfly flight muscle mitochondria a three- to four-fold concentration increase can occur. The initial research on the phosphate transporter was carried out using the swelling technique. Typical experiments are shown in Fig. 4.2. Phosphate efflux can be achieved on the dicarboxylate transporter (see below), this exchange is not inhibited by NEM but is by n-butyl malonate (an inhibitor of the dicarboxylate transporter). The *adenine nucleotide* transporter catalyses a one for one exchange of ADP. This means that the matrix adenine nucleotide concentration remains constant. Coupled mitochondria preferably transport ADP into the matrix and ATP out of the matrix, since ADP is transported into the matrix 10 times faster than ATP. In the presence of an uncoupler the rates of ATP and ADP transport into the matrix are equal. AMP and other nucleotide phosphates, e.g. GTP, are not transported on this carrier.

Table 4.2 Michaelis constants of transporters at $9°C$

Transporter substrate	V_{max} (nmol/min/mg)	K_m (μM)
ADP/ATP	70	12
Phosphate	80	250
Pyruvate	0.6	150
Succinate	50	830
Citrate	23	90
Oxoglutarate	43	46
Glutamate/OH̄	5	4 000

Atractyloside was the first inhibitor of this transporter to be studied in detail, its action on mitochondria in the oxygen electrode was apparently exactly the same as that of oligomycin (Fig. 3.17). It was quickly discovered that atractyloside inhibited the transporter and not the ATPase. There are now three well studied inhibitors of this transporter, atractyloside, carboxyatractyloside and bongkrekic acid, they all act at very low concentrations ($K_{i}s \sim 10^{-8}M$) and they are respectively partially competitive, non-competitive and uncompetitive inhibitors. These inhibitors have enabled very detailed kinetics of the adenine nucleotide transporter to be measured (Table 4.2), and greatly assisted attempts to isolate the molecule involved in the transport process (see later) since the inhibitors will bind strongly and specifically to the transporter. If they are radiolabelled, the purification of the transporter can be followed by monitoring the radiolabel.

In view of the importance of *pyruvate* as a substrate for the tricarboxy-

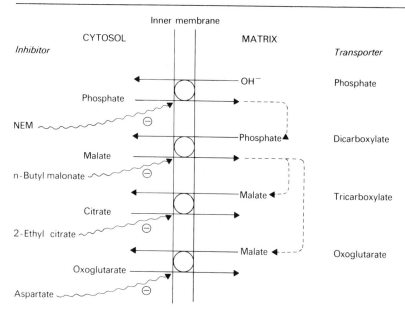

Fig. 4.6   The phosphate transporter is linked to other transporters, i.e. the dicarboxylate, tricarboxylate and oxoglutarate transporters. ⊖ implies inhibition.

lic acid cycle (Ch. 2) it is probable that mitochondria from all tissues also contain a transporter for pyruvate. This transporter is necessary since although pyruvate dehydrogenase is membrane bound (Ch. 2) it requires its substrate to be in the matrix. The presence of a transporter for pyruvate was postulated in 1971 and this was confirmed when a specific inhibitor for this transporter was found. Pyruvate exchanges for hydroxyl ions and using the 'inhibitor-stop' technique with the inhibitor α-cyano-4-hydroxycinnamate; detailed kinetics have been worked out for this transporter (Table 4.2). α-Cyano-4-hydroxycinnamate is a non-competitive inhibitor with a K_i of 6.3 μM. The pyruvate transporter is also inhibited by phenyl pyruvate, which builds up during the disorder, phenylketonuria, it is a competitive inhibitor with a K_i of 1.8 mM. The transporter probably transports other monocarboxylic acids, e.g. acetoacetate, and may be affected by hormones, e.g. glucagon.

It was found from swelling experiments that dicarboxylic acids, e.g. succinate and malate, would not penetrate into the matrix unless there was a little phosphate present (Fig. 4.2(a)). Similarly tricarboxylic acids (citrate, isocitrate and oxoglutarate) would not penetrate unless catalytic amounts of phosphate and malate were present. This indicated that tnree transporters could be linked together (Fig. 4.6). The *dicarboxylate* transporter can catalyse dicarboxylate/dicarboxylate, dicarboxylate/phosphate and phosphate/phosphate exchanges and all these exchanges are competitively inhibited by n-butyl malonate and 2-phenyl succinate. The dicarboxylate anion can be malate, succinate or malonate, but fumarate is not transported. Oxaloacetate can be exchanged for dicarboxylate or phosphate anions, probably on the dicarboxylate transporter, since this exchange is inhibited by n-butyl malonate. The ability to transport oxaloacetate is not as widespread in mammalian mitochondria as that of malate and succinate, e.g. rabbit and pigeon liver mitochondria are not permeable to oxaloacetate whereas rat liver is. Movement of oxaloacetate in pigeon or rabbit liver is probably achieved by transporting malate or citrate which are then converted to oxaloacetate by the action of cytosolic malate dehydrogenase or citrate lyase respectively. The transfer of oxaloacetate may be important for transporting the reducing equivalents out of the mitochondrial matrix (see sect. 4.7).

The *tricarboxylate* transporter can catalyse a tricarboxylate/tricarboxylate exchange and a tricarboxylate/dicarboxylate exchange and these exchanges are inhibited by certain citrate analogues, e.g. 2-ethyl citrate and benzene-1, 2, 3-tricarboxylate. Citrate and isocitrate can be exchanged for citrate, isocitrate, malate or isomalate on this transporter. Phosphoenolpyruvate will exchange with citrate in the mitochondrial matrix and this is inhibited by benzene-1, 2, 3-tricarboxylate, indicating that phosphoenolpyruvate is transported on the tricarboxylate carrier. *Oxoglutarate* is transported on a different carrier to that for citrate and isocitrate, since its transport is not inhibited by 2-ethyl citrate, but is by aspartate and it will exchange with malate or malonate, but not isomalate, whereas citrate and isocitrate will exchange with malate or isomalate, but not malonate. The kinetics of the dicarboxylate and tricarboxylate transporters are shown in Table 4.2.

There are two *glutamate* transporters which have been studied in rat liver mitochondria. One exchanges glutamate for hydroxyl ions and is inhibited by N-ethyl maleimide. The kinetics for this transporter are shown in Table 4.2. The other transporter probably acts exclusively to transport glutamate into the matrix in exchange for aspartate (see sect. 4.7), this carrier is inhibited by glisoxepide and is discussed in section 4.7.1.

As previously stated probably only the adenine nucleotide, phosphate and pyruvate transporters occur in all mitochondria. In fact in blowfly flight muscle mitochondria these are probably the only transporters of those described which are present. This is because the role of flight muscle mitochondria is almost exclusively the production of ATP and this can be achieved with two substrates; pyruvate and α-glycerophosphate, the latter does not require a transporter (Fig. 4.8), as it directly reduces a mitochondrial flavoprotein. The importance of the other transporters in Table 4.1 will be discussed later.

4.5   The biological importance of the transporters

In Table 4.1 the probable biological importance of the mitochondrial transporters is given. The adenine nucleotide and phosphate transporters are involved in ATP synthesis. They transfer ADP and phosphate to the ATPase (Ch. 3) whose active site is on the matrix side of the membrane. Phosphate and ADP also activate certain matrix enzymes, e.g. NAD^+-linked isocitrate dehydrogenase and glutaminase. Mitochondria would have to accumulate phosphate via its transporter since 20—30 mM phosphate is required for maximum activity of NAD^+-linked isocitrate dehydrogenase in some systems, e.g. insect flight muscle mitochondria, and this concentration in the matrix could not be achieved by diffusion. Phosphate is also involved in the net movement of dicarboxylic and tricarboxylic acids.

Malate transport by the dicarboxylate transporter provides the carbon skeleton, in the cytosol, for gluconeogenesis, since malate can be converted to phosphoenolpyruvate by malate dehydrogenase and phosphoenolpyruvate carboxykinase (Ch. 2). The tricarboxylate transporter provides citrate in the cytosol, which can give rise to oxaloacetate or acetyl CoA for gluconeogenesis or fatty acid synthesis respectively. The concentration of citrate in the cytosol, determined by the tricarboxylate transporter, determines the activity of a regulating enzyme of glycolysis, phosphofructokinase. Isocitrate is used in the cytosol to produce NADPH via $NADP^+$-linked isocitrate dehydrogenase, for fatty acid synthesis.

The glutamate/hydroxyl exchange could provide glutamate and therefore NH_3 in the matrix for urea synthesis (Ch. 2). However, the high K_m and low V_{max} values (Table 4.2) indicate that NH_3 may not be produced rapidly enough, via this transporter, for urea synthesis. The production of NH_3 from glutamate dehydrogenase in mitochondria whose membranes have been destroyed is 10 times greater than that in intact mitochondria, indicating that the transport of glutamate across the membrane does exert some control on the rate of NH_3 production.

The glutamate/aspartate exchange can provide aspartate in the cytosol for the urea cycle (Ch. 2) and it can be converted to oxaloacetate by transamination with oxoglutarate and then be another source of the carbon skeleton for gluconeogenesis.

A very important role for the oxoglutarate transporter, glutamate/aspartate exchange and possibly the dicarboxylate transporter is the transfer of reducing equivalents into the mitochondrial matrix (discussed in sect. 4.6).

4.6   The transfer of reducing equivalents into the matrix

The inner mitochondrial membrane is impermeable to NADH, therefore the transfer of the reducing equivalents of NADH into the matrix where

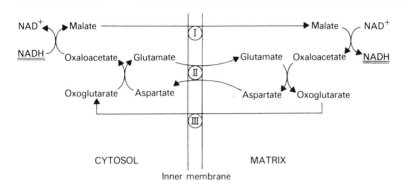

Fig. 4.7   The Borst cycle.
I Dicarboxylate transporter (probably not used as malate can exchange on III),
II Glutamate/aspartate exchange, III oxoglutarate transporter.

oxidation to NAD^+ provides electrons for the respiratory chain is achieved by shuttles. The Borst cycle or shuttle (Fig. 4.7) which operates in most mammalian systems, e.g. liver, heart and brain involves isoenzymes of malate dehydrogenase and aspartate amino transferase in the matrix and cytosol. The NADH is oxidised by malate dehydrogenase and the malate goes into the matrix possibly on the dicarboxylate transporter. The malate is converted to oxaloacetate by matrix malate dehydrogenase with the accompanying reduction of NAD^+. The oxaloacetate in the cytosol is replenished by the transamination of aspartate with oxoglutarate which were formed by the matrix transamination of glutamate and oxaloacetate and then transported out of the matrix on the glutamate/aspartate exchange and oxoglutarate transporter respectively. The oxidation of NADH transferred into the matrix by this cycle gives 3 ATPs. It is probable that the dicarboxylate transporter may not be involved in the Borst cycle since malate could exchange for oxoglutarate on the oxoglutarate transporter only. The glutamate/aspartate exchange (discussed in sect. 4.7.1) is irreversible making the Borst cycle irreversible. Therefore this shuttle can only be used for transferring reducing equivalents into the matrix.

There is another shuttle to transfer reducing equivalents into the matrix, the glycerophosphate shuttle. This is not so important in mammalian mitochondria since α-glycerophosphate is not a very important substrate for mammalian mitochondria, but is very important in insect mitochondria. In this glycerophosphate shuttle (Fig. 4.8) the mitochondrial glycerophosphate dehydrogenase is membrane bound, and so transporters for glycerophosphate or dihydroxyacetone phosphate are unnecessary. NADH in the cytosol is oxidised through the cytosolic dehydrogenase by the conversion of dihydroxyacetone phosphate to glycerophosphate. This

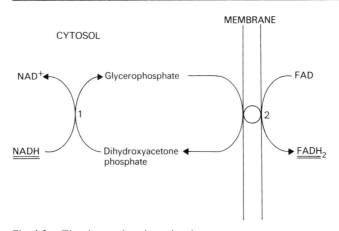

Fig. 4.8 The glycerophosphate shuttle.
1. Cytosolic glycerophosphate dehydrogenase. 2. Mitochondrial glycerophosphate dehydrogenase.

oxidation of one molecule of NADH in the cytosol gives rise to 2 ATPs since the mitochondrial dehydrogenase is flavin-linked (Ch. 3).

4.7 Electrogenic transporters

As previously described the membrane potential across the mitochondrial inner membrane allows some transporters to exchange ions of unlike charge. These transporters are probably unidirectional since the membrane potential is unlikely to be reversed *in vivo*. Four transporters with these characteristics will now be described.

4.7.1 Glutamate/aspartate exchange

The glutamate/aspartate exchange is an integral part of the Borst cycle (Fig. 4.7) and apparently exchanges a negatively charged anion for another negatively charged anion. However, it has been demonstrated that an H^+ ion accompanies the glutamate into the matrix. Therefore this transporter catalyses the electrogenic movement of a neutral molecule (glutamate⁻H^+) into the matrix in exchange for negatively charged aspartate (Fig. 4.5(*d*)). The Borst cycle can therefore probably only move reducing equivalents into the matrix, another mechanism would be required for efflux — possibly a malate/oxaloacetate shuttle in mitochondria where oxaloacetate can be transported on the dicarboxylate transporter (Fig. 4.9). In the presence of uncoupler the glutamate/aspartate exchange, and therefore the Borst cycle, would not operate.

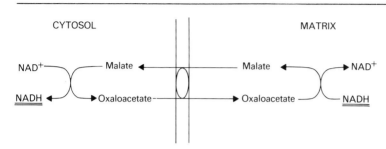

Fig. 4.9 A possible mechanism for the transfer of reducing equivalents out of the matrix.

4.7.2 Ornithine/citrulline transport

As stated in the introduction the site of action of the enzymes of the urea cycle (Ch. 2) necessitates the movement of citrulline and ornithine across the inner mitochondrial membrane. For every molecule of ornithine transported into the matrix one molecule of citrulline is synthesised and transported out. It has been proposed that an electrogenic antiporter may transport ornithine, positively-charged, in exchange for citrulline, neutral at physiological pH. Evidence so far, however, indicates separate transporters for these molecules, that for ornithine being an electrogenic uniporter.

4.7.3 Glutamine/glutamate exchange

In pig kidney mitochondria ammonia for excretion is produced by the action of glutaminase in the matrix, which converts glutamine to glutamate and ammonia. It has been shown that when glutamine is added to isolated mitochondria in the absence of ADP, glutamine is almost quantitatively converted to glutamate which accumulates outside the mitochondria. In the presence of ADP the glutamate is transported into the matrix on the glutamate/aspartate exchange and transamination with oxaloacetate produces the aspartate (Fig. 4.10). Since at physiological pH glutamate is negatively charged and glutamine is neutral, an electrogenic antiporter, which favours the production of NH_3 in the matrix, probably operates in pig heart mitochondria.

Glutaminase is very strongly inhibited by glutamate, therefore this exchange reduces intramitochondrial glutamate allowing the enzyme to deaminate glutamine.

4.7.4 Adenine nucleotide exchange

At physiological pH the adenine nucleotides are charged and the transporter therefore exchanges ADP^{3-} for ATP^{4-}. This overall transfer of a negative charge out of the matrix is possible because of the membrane

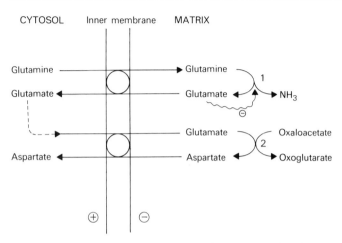

Fig. 4.10   The operation of the glutamine/glutamate exchange.
1. Glutaminase. 2. Aspartate amino transferase.

potential, negative on the matrix side of the membrane. It has been shown that added ADP does not mix rapidly with intramitochondrial ADP when mitochondria are in State III. This would imply that under these conditions the transporter can donate ADP directly to the F_1-ATPase in exchange for ATP.

4.8   Carnitine – the transport of fatty acyl CoA

Fatty acyl CoA cannot cross the mitochondrial inner membrane to reach the enzymes of β-oxidation in the matrix. The conversion of fatty acyl CoA to fatty acyl carnitine allows this movement into the matrix. In heart mitochondria a carnitine/acyl carnitine exchange has been demonstrated in the membrane. It enables fatty acyl carnitine to enter the matrix where it is converted to fatty acyl CoA by carnitine acyl transferase (a membrane-bound enzyme, which requires its substrates to be in the matrix). The carnitine formed in this reaction is transported into the cytosol in exchange for more fatty acyl carnitine. Substrate analogues, e.g. ω-trimethyl amino acyl carnitines, competitively inhibit the carrier.

In rat liver mitochondria it was found that carnitine would not cross the membrane. Since the mitochondrial transferase was probably membrane bound it was postulated that fatty acyl carnitine reacts with the enzyme at the outerface of the membrane, CoA reacts from the inside giving fatty acyl CoA inside the matrix and carnitine in the cytosol. This is an example of a type of transporter called a group translocase (Fig. 4.11).

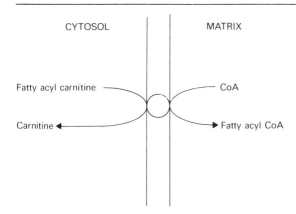

CYTOSOL   MATRIX

Fatty acyl carnitine ———————— CoA

Carnitine ◄———————————————► Fatty acyl CoA

Fig. 4.11   A group translocase.

This mechanism would remove the necessity for a carrier for carnitine, but conclusive evidence for such a mechanism does not exist.

4.9   Transport of cations

4.9.1 Monovalent cations

The inner mitochondrial membrane is impermeable to all monovalent cations, e.g. K^+, Na^+ and H^+, however certain antibiotics, e.g. valinomycin, nigericin and gramicidin allow these cations to cross the membrane. These antibiotics are called ionophores because they are able to 'dissolve' into the phospholipid membrane as the outside of the molecule is hydrophobic. The structure of valinomycin is shown in Fig. 4.12. Cations are able to bind by coordination to the hydrophilic interior of the molecules, therefore the ionophore can carry the cation across the membrane.

Valinomycin is a neutral molecule, therefore its complex with K^+, Rb^+ or Cs^+ will be positively charged. The addition of valinomycin to mitochondria in a potassium-containing medium would make the inner membrane permeable to K^+. If rotenone is present and the external K^+ concentration is higher than that in the matrix then valinomycin will bring about an exchange of K^+ across the membrane, however, there will be no *net* movement of K^+. The addition of uncoupler will make the membrane permeable to H^+ and then there can be *net* movement of K^+ into the matrix.

Nigericin is negatively charged, therefore a nigericin-K^+ complex will be neutral. Nigericin can also complex with hydrogen ions, therefore uncoupler is not needed to allow *net* K^+ penetration into the matrix.

107

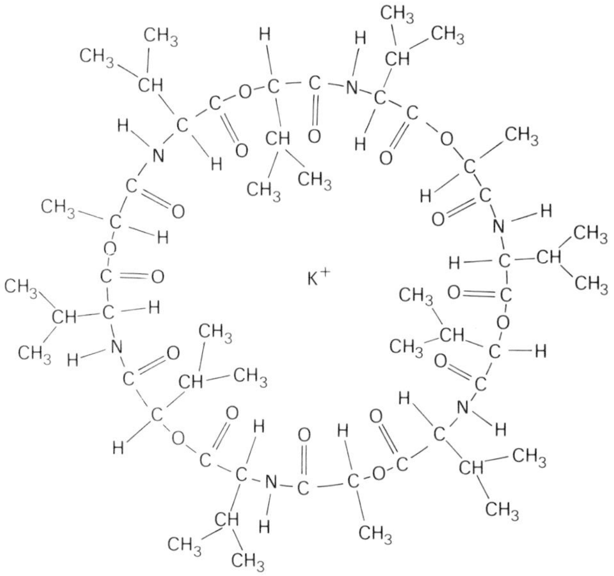

Fig. 4.12   The structure of valinomycin.

Nigericin acts as a carrier, exchanging K^+ for H^+. Nigericin and valinomycin preferentially bind K^+, but gramicidin will bind Na^+ equally well.

4.9.2 Divalent cations

The transport of Ca^{++} into the mitochondrial matrix has been studied in detail by Carafoli and co-workers. Mitochondria can accumulate Ca^{++} against a concentration gradient; phosphate is taken up simultaneously. The energy for this accumulation can be provided by electron flow or ATP hydrolysis. When it is supported by electron flow, Ca^{++} accumulation is inhibited by respiratory chain inhibitors (antimycin A, rotenone) and uncoupler (DNP and FCCP, Ch. 3) and when it is supported by ATP hydrolysis it is inhibited by oligomycin and uncouplers. Mn^{++} and Sr^{++} can also be accumulated in this manner, but *not* Mg^{++}. Electron flow cannot support both Ca^{++} accumulation and ATP synthesis, if ADP and Ca^{++} are present then Ca^{++} will be accumulated before any ATP synthesis occurs. For every two electrons passing down the respiratory chain from NADH to O_2 approximately $6 \, Ca^{++}$ can be accumulated. This accumulation is inhibited by low concentrations of ruthenium red and La^{3+} and other rare earth elements.

Ca^{++} is a very important metabolic regulator in the cell. It activates the reversible contraction of muscle and regulates certain enzymes located in important metabolic pathways, e.g. pyruvate, isocitrate and glycero-phosphate dehydrogenases. The importance of Ca^{++} to the cell has stimu-lated detailed studies of Ca^{++} accumulation and binding by mitochondria. Three classes of Ca^{++} binding site have been described for mitochondria, they are divided into high and low affinity binding sites according to their dissociation constants (K_d). The low affinity binding sites $(K_d \sim 100~\mu M)$ will bind 50–70 nmol Ca^{++}/mg of protein in the absence of energy. The binding of Ca^{++} to these sites is only inhibited by high concentrations of ruthenium red and La^{3+}, much higher concentrations than those required to inhibit Ca^{++} accumulation. These sites occur on the outer as well as the inner mitochondrial membrane and the Ca^{++} is probably binding to phos-pholipids.

There are two types of high affinity Ca^{++} binding sites on mitochondria. One is believed to be associated with α-glycerophosphate dehydrogenase, an enzyme whose active site is on the outer site of the inner membrane (Fig. 4.8). The K_m for glycerophosphate is lowered by increased Ca^{++} concentrations. The binding of Ca^{++} to these sites is not inhibited by ruthenium red. The other high affinity binding sites $(K_d~0.1-1~\mu M)$ can bind 1 nmol Ca^{++}/mg of protein and are believed to be associated with the accumulation of Ca^{++} by mitochondria. They show all the characteristics of energy-linked uptake of Ca^{++} including inhibition by low concentrations of ruthenium red and La^{3+}. In most mitochondria if these high affinity binding sites are absent then there is no accumulation of Ca^{++} (blowfly flight muscle mitochondria are an exception to this rule). This would imply that binding to these high affinity sites is the preliminary process to the transport of Ca^{++} into the matrix.

How Ca^{++} can be released from mitochondria is not well understood. It has been shown that prostaglandin E_1 can release Ca^{++} from the matrix of rat liver mitochondria, and cyclic AMP may release Ca^{++} from adrenal cortex mitochondria. Recently a Na^+/Ca^{++} exchange has been demon-strated in heart mitochondria, this allows Ca^{++} out of the matrix in exchange for 3 Na^+. Although Na^+ is the preferred cation to exchange with Ca^{++}, Li^+ can also be used, but K^+ will not exchange with Ca^{++}.

The importance of calcium accumulation by mitochondria is unknown, although it has been postulated that it may be important for releasing and storing Ca^{++} and thereby controlling metabolism. It has also been postu-lated that the precipitation of calcium phosphate in the matrix may be important in bone formation.

4.10 Isolation of proteins associated with transport

The high specificity of the transport systems described in this chapter has led to the proposal that the carriers are specific proteins in the inner membrane. The next step in the investigation of the carriers is therefore to isolate the proteins concerned and reconstitute the transport process in an artificial phospholipid membrane.

So far attempts have been made to isolate the adenine nucleotide, the glutamate/hydroxyl and the Ca^{++} transporters. The methods involved include treatment of mitochondria with hypotonic media, sonication and detergents followed by purification by dialysis and electrophoresis. The transporter is often labelled with a specific inhibitor before isolation is attempted.

A proteolipid has been isolated from pig heart mitochondria, which shows many characteristics of the glutamate/hydroxyl exchange. It is insoluble in water and contains a lot of phospholipid especially cardiolipin. The proteolipid has a high affinity for glutamate, K_d 62 μM, and NEM inhibits the binding of glutamate. It also has no enzymic properties towards glutamate and promotes glutamate entry into liposomes.

A glycoprotein has been isolated from mitochondrial membranes that will bind Ca^{++} at high and low affinity sites and this binding is inhibited by La^{3+} and ruthenium red. This Ca^{++} binding glycoprotein contains variable amounts of phospholipid. Attempts have been made to incorporate this protein into a phospholipid membrane, but transport in this reconstituted system has as yet not been demonstrated.

A protein has been isolated that is believed to be the adenine nucleotide transporter, since the mitochondria were pretreated with radiolabelled carboxyatractyloside and the labelled protein was isolated. The protein appears to exist in two states, the 'm' state which will bind bongkrekic acid and the 'c' state which will bind carboxyatractyloside and atractyloside. This is consistent with findings in whole mitochondria. The 'm' state is when the transporter is facing into the matrix and the 'c' state when it faces into the intermembrane space. The 'c' state can only be converted to the 'm' state when ADP is present. When bongkrekic acid binds to the 'm' state the protein cannot alter configuration and ADP is held bound to the transporter.

The isolated protein has a molecular weight of 29 000 and may exist as a dimer in the mitochondrial membrane and is probably closely associated with the ATPase.

Suggested further reading

Reviews
CHAPPELL, J. B. (1968) Systems for the transport of substrates into mitochondria, *Brit. Med. Bull.*, **24**, 150—7.
KLINGENBERG, M. (1970) Metabolite transport in mitochondria: An example for intracellular membrane function, *Essays in Biochem.*, **6**, 119—60.
BROUWER, A., SMITS, G. G., TAS, J., MEIJER, A. J. and TAGER, J. M. (1973) Substrate anion transport in mitochondria, *Biochimie*, **55**, 717—25.
CARAFOLI, E. (1973) The transport of calcium by mitochondria: Problems and perspectives, *Biochimie*, **55**, 755—62.

MEIJER, A. J. and VAN DAM, K. (1974) The metabolic significance of anion transport in mitochondria, *Biochim. Biophys. Acta*, **346**, 213–44.

Methods
CHAPPELL, J. B. and CROFTS, A. R. (1966) Ion transport and reversible volume changes of isolated mitochondria, in *Regulation of Metabolic Processes in Mitochondria*, pp. 293–314, eds. J. M. Tager, *et al.* Elsevier Publishing Co., Amsterdam.
CHAPPELL, J. B., HENDERSON, P. J. F., McGIVAN, J. D. and ROBINSON, B. H. (1968) The effect of drugs on mitochondrial function, in *The Interaction of Drugs and Subcellular Components and Animal Cells*, pp. 71–95, ed. P. N. Campbell. J and A Churchill Ltd., London.
HELDT, H. W., KLINGENBERG, M. and MILOVANCEV, M. (1972) Differences between the ATP/ADP ratios in the mitochondrial matrix and the extramitochondrial space, *Eur. J. Biochem.*, **30**, 434–40.
PFAFF, E. and KLINGENBERG, M. (1968) Adenine nucleotide translocation in mitochondria, I Specificity and control, *Eur. J. Biochem.*, **6**, 66–79.

Kinetics and isolation of transporters
BRADFORD, N. M. and McGIVAN, J. D. (1973) Quantitative characteristics of glutamate transport in rat liver mitochondria, *Biochem. J.*, **134**, 1023–9.
CROMPTON, M. and CHAPPELL, J. B. (1973) Transport of glutamine and glutamate in kidney mitochondria in relation to glutamine deamination, *Biochem. J.*, **132**, 35–46.
HALESTRAP, A. P. (1975) The mitochondrial pyruvate transporter. Kinetics and specificity for substrates and inhibitors, *Biochem. J.*, **148**, 85–96.
JULLIARD, J. H. and GAUTHERON, D. C. (1973) High glutamate affinity proteolipid from pig heart mitochondria. Is it a component of a glutamate translocator? *FEBS Letts.*, **37**, 10–16.
LaNOUE, K. F. and TISCHLER, M. C. (1974) Electrogenic characteristics of the mitochondrial glutamate/aspartate antiporter, *J. Biol. Chem.*, **249**, 7522–8.
PALMIERI, F., QUAGLIARIELLO, E. and KLINGENBERG, M. (1972*a*) Kinetics and specificity of the oxoglutarate carrier in rat liver mitochondria, *Eur. J. Biochem.*, **29**, 408–16.
PALMIERI, F., STIPANI, I., QUAGLIARIELLO, E. and KLINGENBERG, M. (1972*b*) Kinetic study of the tricarboxylate carrier in rat liver mitochondria, *Eur. J. Biochem.*, **26**, 587–94.
PFAFF, E., HELDT, H. W. and KLINGENBERG, M. (1969) Adenine nucleotide translocation in mitochondria. Kinetics of the adenine nucleotide exchange, *Eur. J. Biochem.*, **10**, 484–93.
QUAGLIARIELLO, E., PALMIERI, F., PREZIOSO, G. and KLINGENBERG, M. (1969) Kinetics of succinate uptake by rat liver mitochondria, *FEBS Letts.*, **4**, 251–4.
RAMSAY, R. R. and TUBBS, P. K. (1975) The mechanism of fatty acid uptake by heart mitochondria: An acylcarnitine–carnitine exchange, *FEBS Letts.*, **54**, 21–5.
RICCIO, P., AQUILA, H. and KLINGENBERG, M. (1975) Solubilization of the carboxyatractyloside binding protein from mitochondria, *FEBS Letts.*, **56**, 129–32.
RICCIO, P., AQUILA, H. and KLINGENBERG, M. (1975) Purification of the carboxyatractyloside binding protein from mitochondria, *FEBS Letts.*, **56**, 133–8.

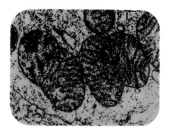

Chapter 5

The assembly of mitochondria

5.1　Introduction

The synthesis of new mitochondrial components occurs throughout the growth cycle of cells and tissues. This is clearly essential to maintain a constant level of mitochondrial activity in a growing cell population. In the previous chapters the complexity of mitochondrial structure and function has been detailed. In recent years considerable interest has developed in the problem of how the assembly of this complex structure is organised. There have been several stimuli to interest in this problem. In the first place biochemists have achieved an understanding of the fundamental processes involved in the synthesis of the simpler biomolecules and the various classes of macromolecules, and considerable progress is being made towards establishing the control processes which modulate their syntheses. Investigation of the mechanism of assembly of a complex organelle such as the mitochondrion represents one of the logical next challenges to the biochemist who is interested in how a cell achieves its total structure and organisation. Secondly, the difficulties which have been encountered in establishing the mechanism of oxidative phosphorylation have also stimulated the study of mitochondrial assembly. A full understanding of mitochondrial assembly will undoubtedly be of immense value to understanding the detailed structure and function of the organelle. A third stimulus came from the discoveries of mitochondrial DNA and protein synthesis. Molecular biologists were quick to develop an interest in the properties and functions of these. A combined attack from these various viewpoints is beginning to reveal a composite picture of the synthesis of the components of mitochondria, in particular those of the inner membrane, the integration and regulation of their synthesis and their genetic origins.

5.2   Cellular origin of mitochondria

Despite the considerable advances which have been made in understanding the biochemical aspects of mitochondrial assembly, there is still a degree of uncertainty about the answer to the problem of the cellular origin of mitochondria. Prior to the discovery that mitochondria contained DNA as well as a protein-synthesising system distinct from that associated with the cytosol and endoplasmic reticulum it was commonly believed that mitochondria originated from invagination or evagination of other cellular membranes. However, the discovery of mitochondrial DNA and protein synthesis in conjunction with the distinctive lipid composition of the mitochondrial inner membrane suggested that the mitochondria have a degree of autonomy which, although somewhat exaggerated in the popular literature, suggests at least that mitochondrial continuity from generation to generation should be the case. There is a good deal of research which supports this hypothesis. Perhaps the most elegant demonstration was that of Manton referred to in section 1.1. The unicellular flagellate *Chromulina* was clearly shown to possess a single mitochondrion which undergoes binary fission during the cell-division process. Unfortunately such clear cut evidence is not available for many organisms. A striking example in higher organisms is shown in the electron micrographs of mitochondria of developing fat body from the larva of the moth *Calpodes* produced by Larsen which show the presence of a septum at the equatorial plane as though the mitochondria are in a process of division (Fig. 5.1). Recent work also mentioned in Chapter 1 suggests that the mitochondria of an individual cell represent a spatial or temporal continuum within the cell fragmenting or coalescing in a somewhat random fashion and implies that mitochondria probably grow by accretion of newly synthesised materials, and fragment in either an organised fashion as in the case of *Chromulina* or in a more unpredictable fashion such as in *Saccharomyces cerevisiae*.

A biochemical approach to the study of mitochondrial origin was employed by Luck using the fungus *Neurospora crassa*. Luck took advantage of the fact that the base choline is concentrated in the phospholipids of the mitochondrial inner membrane. Luck obtained choline-requiring mutants of *Neurospora* which have to be provided with choline in the growth medium in order to grow. The mutant strain was grown in a medium containing ^{3}H-labelled choline and then transferred to medium containing unlabelled choline. After three mass doublings in the presence of unlabelled medium the cells were fragmented and the mitochondria were isolated and spread onto photographic plates which were sensitive to the β-particles emitted by the ^{3}H. After developing the plates the distribution of label between the mitochondria was estimated.

Luck reasoned that if new mitochondria arose by division of pre-existing mitochondria then all of the mitochondria would be evenly labelled after three rounds of growth in unlabelled medium, whereas if

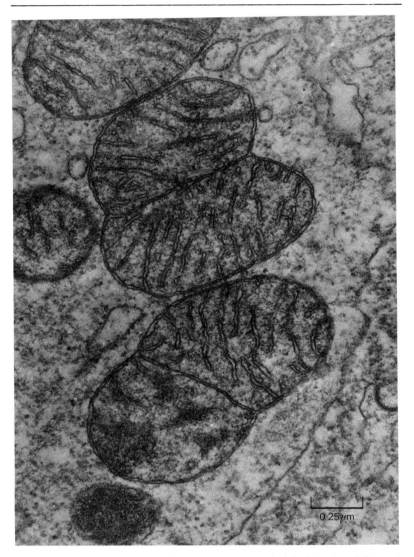

Fig. 5.1 Dividing mitochondria from a newly emerged adult *Calpodes ethlius.*
Osmium tetroxide fixed, uranyl acetate stained. (Courtesy of Dr W. J. Larsen.)

new mitochondria were assembled *de novo* or by modification of vesicles
nipped off from some other cellular membranes then one might expect to
find two populations of mitochondria — those containing labelled choline
(which had been carried forward from the period of labelling) and un-

labelled mitochondria (developed during the period when the culture was growing in the presence of only unlabelled choline). In fact the mitochondria were evenly labelled indicating the origin of new mitochondria from pre-existing labelled mitochondria. Despite the elegance of this biochemical approach, as Luck himself was aware, there are other interpretations. It is possible that there was a turnover of choline in the cell so that it was constantly released and reincorporated in the mitochondria. This would lead to randomisation of the tritium throughout the mitochondrial population. Any coalescence and fragmentation of the mitochondria of the type mentioned earlier would have the same effect.

In summary, it seems likely that all new mitochondrial structure (or at least inner membrane structure) is achieved by addition to pre-existing mitochondria of newly synthesised mitochondrial components. New mitochondria are, in all probability, invariably formed by a division process which may be more or less precise depending on the organism or tissue in question.

5.3 Mitochondrial nucleic acids and protein synthesis

The first clear demonstration that mitochondria contained a DNA fraction distinct from that of the nucleus came in 1963 from Nass and Nass who were investigating the fine structure of developing chick embryos. They observed that in electron micrographs there appeared uranyl acetate-stained fibres within the matrix of mitochondria. The fibres were not present after DNAase (deoxyribonuclease) treatment and were thus considered to be DNA.

A more clear cut demonstration came in 1965 when Mahler and other workers extracted DNA from yeast cells and subjected this to analytical ultracentrifugation in a caesium chloride gradient. At equilibrium the DNA separated into two distinct bands with buoyant densities of 1.683 and 1.699 g cm^{-3} (Fig. 5.2(a)). The 1.683 g cm^{-3} band represented about 15 per cent of the total DNA. Isolation of a mitochondrial fraction from the yeast followed by extraction of DNA from this, resulted in a considerably decreased 1.699 g cm^{-3} component (Fig. 5.2(b)), while the 1.683 g cm^{-3} component was maintained. It was clear that the major, high density component is the nuclear DNA and the minor component is mitochondrial DNA. Subsequently mitochondrial DNAs have been identified in a wide range of organisms. The difference in density is not usually as great as is seen in yeast (Table 5.1) and the quantity of DNA is not always as high as observed in yeast.

Electron microscopic examination of mitochondrial DNAs spread on photographic plates showed that in most cases mitochondrial DNA is circular. In animal cells, with the exception of some protozoa the molecu-

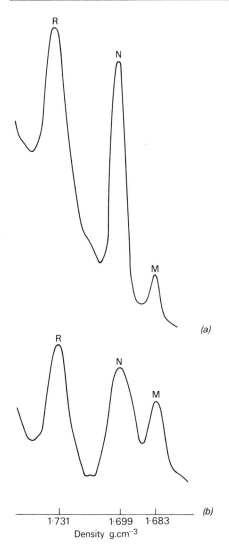

Fig. 5.2 Mitochondrial DNA in *Saccharomyces cerevisiae*.
(a) DNA extracted from whole cells, and analysed on a caesium chloride gradient in the presence of reference DNA (from *Micrococcus luteus*). Distribution of DNA within the gradient was established by photographic recording of ultraviolet light absorption. Photographic negatives were scanned with a microdensitometer.
R = reference DNA; N = nuclear DNA; M = mitochondrial DNA. (b) as (a) except that DNA was extracted from partly purified mitochondria.

Table 5.1   Mitochondrial DNAs from various species

Species	Size (μm)	Density in neutral CsCl g cm^{-3}	
		Nuclear DNA	Mitochondrial DNA
Yeasts			
Saccharomyces cerevisiae	27 (circular)	1.699	1.683
Kluyveromyces lactis	11.4 (circular)	1.699	1.692
Schizosaccharomyces pombe	6.0 (circular)	1.695	1.689
Protozoa			
Paramecium aurelia	18 (?)	1.689	1.702
Tetrahymena pyriformis	17.6 (linear)	1.690	1.684
Algae			
Euglena gracilis	5.0 (?)	1.707	1.690
Higher plants			
Pisum sativum (pea)	32.0 (circular)	1.692	1.706
Insects			
Housefly (*Musca domestica*)	5.2 (circular)	—	—
Drosophila melanogaster	—	1.686	1.696
Fish			
Carp	5.4 (circular)	1.697	1.703
Amphibia			
Rana pipiens (frog)	5.9 (circular)	1.702	1.702
Xenopus laevis	5.7 (circular)	1.702	1.704
Birds			
Chick	5.1—5.4 (circular)	1.701	1.708
Duck	5.1 (circular)	1.700	1.711
Mammals			
Rat	5.1—5.4 (circular)	1.703	1.701
Cow	5.1—5.3 (circular)	1.704	1.702
Monkey	5.5 (circular)	—	—
Man	4.8—5.3 (circular)	1.695	1.705

lar circumference is usually approximately 5 μm (Fig. 5.3(*a*)). In bakers' yeast most of the molecules are linear and of rather variable length. However, a small proportion of the molecules observed are circular, supercoiled and have a circumference of 27 μm (Fig. 5.3(*b*)). This is presumed to be the size of the mitochondrial DNA molecule *in vivo*.

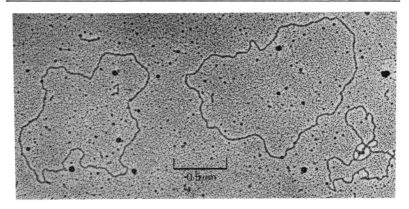

Fig. 5.3(*a*) Rat liver mitochondrial DNA. Mitochondrial DNA was prepared for electron microscopy by the DNA-protein monolayer technique and rotary shadowed by platinum/iridium. (Courtesy of Dr M. M. K. Nass.)

Higher plant mitochondrial DNAs are poorly characterised but are thought to be somewhat larger than animal mitochondrial DNAs. Cells seem to possess many molecules of mitochondrial DNA. In bakers' yeast for example, there are usually 50–100 molecules per cell. The precise number of molecules is related to the ploidy level (the number of full sets of chromosomes per cell) there being approximately twice as many in diploid as in haploid cells, thus maintaining a constant proportion of mitochondrial DNA relative to nuclear DNA. Rat liver cells contain upward of 1 000 mitochondria per cell averaging 4–5 mitochondrial DNA circles per mitochondrion. The renaturation rate of heat denatured mitochondrial DNA is very rapid and it is calculated from such determinations that the mitochondrial DNA circles are probably all identical within a single cell. This can give us an upper limit for the amount of genetic information which may be encoded in the mitochondrial DNA molecule. If we consider animal cell mitochondrial DNA (5 μm circles) it is easily calculated that this is equivalent to about 1.5×10^4 base pairs (distance between base pairs $= 3.4$ Å). This could specify 5×10^3 amino acids (30–50 small proteins). This must obviously be an upper limit as at least part of the genome must be taken up by gene spacers and control elements. The bakers' yeast mitochondrial genome (probably 27 μm) could code for approximately 50 small proteins in theory. Bernardi has shown that the very low density (1.683 g cm^{-3}) is in part due to the presence of genetically meaningless spacers containing almost exclusively adenine and thymine at intervals in the molecule. Despite the fact that they represent almost half the mitochondrial genome in bakers' yeast their function is unknown. However, it does mean that the number of small proteins that

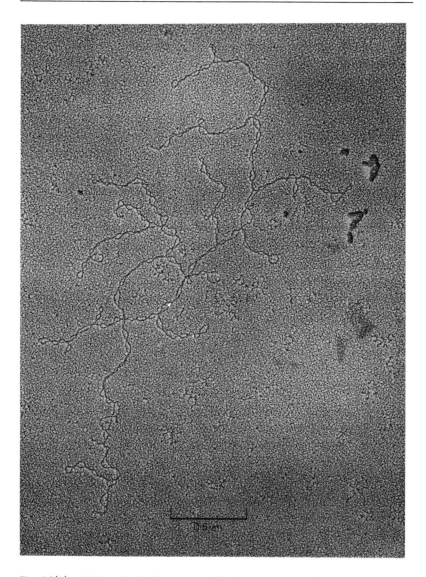

Fig. 5.3(*b***)**   DNA molecule from osmotically lysed yeast mitochondria. Mitochondrial DNA was prepared for electron microscopy by the DNA—protein monolayer technique and rotary shadowed with platinum. (Courtesy of Dr. A. J. Van Bruggen.)

can be coded by mitochondrial DNA in bakers' yeast is somewhat less than that suggested by the contour length of the mitochondrial DNA molecule.

What emerges from all of these observations is that the amount of genetic information available in mitochondrial DNA falls far short of that needed to code for all mitochondrial proteins and the major contribution must presumably be made by nuclear DNA. Clearly the mitochondria cannot really be considered to be autonomous as has been suggested in some popular accounts of mitochondrial biogenesis.

If the mitochondrial DNA is to have a genetic role, it is necessary to establish that it is capable of being replicated and transcribed. Both DNA polymerase and DNA-dependent RNA polymerase have been isolated from mitochondria from a number of species and have been shown to be physically and enzymologically distinguishable from their nuclear counterparts. There is abundant electron micrographic evidence for the replication of mitochondrial DNA, apparent replication intermediates having been observed from a number of cell types (Fig. 5.4). Electron migrographs have also been obtained showing Hela cell mitochondrial DNA apparently undergoing transcription.

Mitochondria also contain the machinery for translation of messenger RNA into proteins. In particular they contain within the matrix, ribosomes (mitoribosomes) which are physically distinguishable from those of the endoplasmic reticulum and cytosol. These seem to be largely bound to the matrix facing surface of the inner membrane. Non-mitochondrial eukaryotic ribosomes have almost invariably a sedimentation coefficient of $80\,S$ and are composed of large $(60\,S)$ and small $(40\,S)$ subunits. Mitochondrial ribosomes are usually (but not always) smaller than this and there seems to be a good deal of interspecies variation. Only rarely are the sedimentation properties of mitochondrial ribosomes the same as those usually associated with prokaryotes ($70\,S$ ribosome; $50\,S$ and $30\,S$ subunits). There is also considerable variation in the sedimentation coefficients of mitoribosomal RNAs compared with either extramitochondrial or prokaryotic ribosomes. Like other ribosomes, mitoribosomes each contain two high molecular weight RNAs one associated with each subunit. However, up till now, no equivalent of the large subunit $5\,S$ RNA found in both eukaryotic cytosolic and prokaryotic ribosomes, and essential to the assembly of the subunit, has been discovered.

If there are important structural differences between mitoribosomes and eukaryotic cytosolic ribosomes (cytoribosomes) or prokaryotic ribosomes they nevertheless exhibit certain similarities. There is a good correlation between the base compositon of mitoribosomes and that of their extramitochondrial equivalents. As a general rule mitoribosomal RNA contains about 1.5 times the proportion of guanine and cytosine as the equivalent cytoribosomal RNA. Some of the initiation and elongation factors active in mitochondrial protein synthesis are also active in bacterial (but not in cytoribosomal) protein synthesising systems. Furthermore,

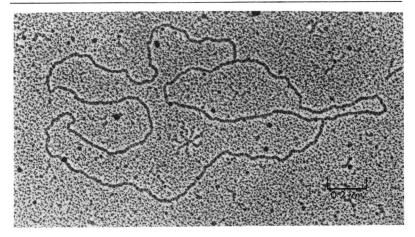

Fig. 5.4   Replication intermediate of rat liver mitochondrial DNA. Mitochondrial DNA was prepared for electron microscopy by the DNA—protein monolayer technique. Grids were rotary shadowed with platinum/palladium. (Courtesy of Dr D. R. Wolstenholme.)

mitochondrial protein synthesis is sensitive to a number of inhibitors of prokaryotic protein synthesis (e.g. erythromycin and chloramphenicol) but insensitive to a number of inhibitors of cytoribosomal protein synthesis (e.g. cycloheximide or emetine).

Mitochondria have been shown to possess the two other major classes of RNA essential for protein synthesis, messenger RNA (mRNA) and transfer RNA (tRNA). Messenger RNA has been demonstrated as a labelled product following pulse exposure to a radioactively-labelled precursor (usually adenine or uracil). mRNA is recognised as being a very small fraction of total mitochondrial RNA and having a very heterogeneous distribution of base composition and molecular weight. Mitochondrial mRNA molecules have been shown to contain a polyriboadenylic acid (polyA) sequence at the 3′ hydroxyl end of the molecule. This is similar to the situation found in messenger RNAs of nuclear origin (although the polyA sequence is usually longer in the nuclear mRNA). Prokaryotic mRNAs do not have this polyA tail.

There is a growing catalogue of transfer RNAs and their corresponding aminoacyl-tRNA-synthetases in mitochondria from various sources. In most cases these can be readily differentiated from their counterparts in the cytosol. In summary, it is now clearly established that mitochondria contain DNA which is capable of replication and a complete machinery for the expression of the information included within the DNA molecule.

5.4   Mitochondrial mutants

One of the techniques which has contributed a great deal to the understanding of mechanisms of assembly of simple biomolecules and macromolecules, has been the study of mutants defective in some aspect of the
assembly process. Similarly, mutants have played a significant role in
starting to unravel the problem of mitochondrial assembly. Early interest
was stimulated by the discovery by Ephrussi and his co-workers in 1949,
of mutants of bakers' yeast (*Saccharomyces cerevisiae*) which formed
colonies more slowly than normal on agar plates containing glucose as a
carbon source. He referred to these mutants as '*petite colonie*' mutants.
They are now usually referred to as *petite* mutants. The mutants were
subsequently shown to have lost the ability to make a normal ATP
synthesising system in the mitochondria. They would grow if provided
with a fermentable carbon source (such as glucose or sucrose) – obtaining
their ATP requirements by alcoholic fermentation – but not if the sole
carbon source was non-fermentable (such as glycerol or ethanol). Because
colony size does not always give a clear indication of whether or not it is
petite, it is now usual to diagnose *petite* colonies by overlaying them with
agar containing 2,3,5-triphenyltetrazolium chloride which is reduced by
grande (wild-type) colonies to a red compound, but not by *petite* colonies.

It is this unique capability to form *petite* mutants along with its ability
to grow rapidly on a defined medium which has led to the widespread use
of *Saccharomyces cerevisiae* as an experimental organism for the study of
the mitochondrial assembly.

A number of other unusual observations on *petite* mutants suggested
that something rather different from a normal mutation was occurring. In
the first place *grande* strains spontaneously mutated to *petite* at an
incredibly high rate (0.1 to 1 per cent generation). The DNA-intercalating
drug acriflavine was found to be capable of inducing practically 100 per
cent mutation in a few generations of growth (ethidium bromide is now
known to be even more effective). Furthermore, when *petite* mutants were
subsequently cultured no revertant *grandes* could ever be picked up.

When Ephrussi and his colleagues looked at the genetics of these
mutants they were further surprised. Before summarising their genetic
data, it may be worth while to outline the life cycle of *Saccharomyces
cerevisiae* (Fig. 5.5). The yeast may exist in either a haploid (*n*) or diploid
(*2n*) form. There are two mating types *a* and α which are physiologically
but not morphologically distinguishable. Fusion will occur between an *a*
strain and one of an α strain. The zygote formed will give rise to diploid
cells by budding. The diploid cells may be induced to sporulate (form
spores) when they will undergo meiosis (reduction division) to produce an
ascus (spore sac) containing four haploid ascospores 2 of *a* and 2 of α
mating types.

Ephrussi was able to divide the *petites* he isolated into three classes

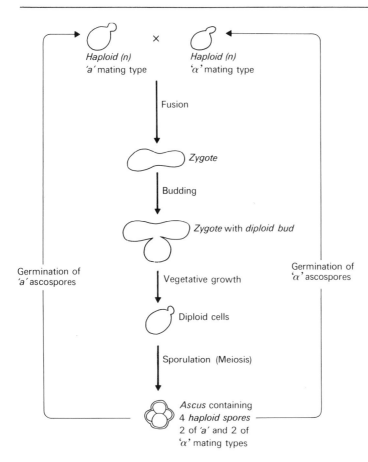

Fig. 5.5   Life cycle of *Saccharomyces cerevisiae.*

depending on their behaviour in a simple cross of this nature (Table 5.2). In all crosses a *petite* strain was mated with a haploid *grande* of opposite mating type. The segregation of a representative nuclear gene at meiosis was always used as a control. Behaviour of the chromosomes at meiosis results in the appearance of the particular allele of a nuclear chromosomal gene from one of the parent strains in two of the ascospores of an ascus and the allele of that gene from the other parent in the other two ascospores of the ascus. The segregation of the *a* and α alleles of the mating type gene referred to above illustrates this principle. The first of the three classes of *petite* mutants (segregational or nuclear *petites*) occurred very rarely and segregation in the ascus was 2:2 *grande:petite* as

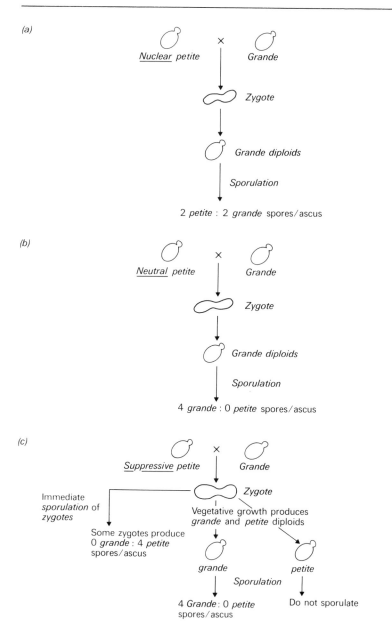

Fig. 5.6 Genetics of *petite* mutants. (*a*) nuclear *petites*; (*b*) neutral *petites*; (*c*) suppressive *petites*.

Table 5.2   Properties of *petite* mutants

Type of *petite*	Occurrence	Phenotype of diploids produced in *petites* × *grandes*	Segregation of *grande:petite* after sporulation of diploids
Nuclear	Rare	*Grande*	2:2
Neutral	Common	*Grande*	4:0
Suppressive	Common	*Grande*	4:0
		petite	no sporulation

expected with a nuclear gene mutation (Fig. 5.6(a)). The second class of *petites* (neutral *petites*) occurred more commonly and like the nuclear *petites* produced all *grande* diploids (Fig. 5.6(b)). On sporulation these did not segregate *grande* and *petite* cells as did the nuclear *petites*, but gave all *grande* cells. Repeated backcrosses of the resulting haploid *grandes* with the original parent *petite* strain failed to elicit any *petite* progeny so that it looked very much as if the genetic determinant for the *grande/petite* phenotype was not a normal chromosomal gene (or collection of chromosomal genes).

The third type of *petite* (suppressive *petite*) which occurred most commonly was also the strangest (Fig. 5.6(c)). In this case the diploid progeny could be either *grande* or *petite*, and so the *petite* characteristic is somehow able to suppress the *grande* characteristic in a proportion of the diploids. The percentage of *petite* diploids (degree of suppressiveness) was characteristic for a particular suppressive *petite* strain. On sporulation of *grande* diploids a 4:0 ratio of *grande* to *petite* was obtained. The *petite* diploids were unable to sporulate because sporulation requires a functional mitochondrial ATP synthesising system. If, however, zygotes from such a cross were sporulated immediately following mating a number of asci were shown to have a 0:4 *grande:petite* ratio. (In these cases it is likely that the mitochondrial ATP synthesising apparatus from the original *grande* parent survived long enough to provide for sporulation.)

These results indicated a number of things:
1. There appeared to be a non-chromosomal genetic determinant, mutation or loss of which gives rise to the *petite* phenotype.
2. Mutation or loss of the determinant is either recessive (in neutral *petites*) or partially dominant (in suppressive *petites*).
3. At least one nuclear gene is necessary to maintain the non-chromosomal determinant. Mutation in the nuclear gene gives rise to segregational *petites* (several such genes are now known).

Further light was shed on *petite* mutants when biochemical studies on mitochondrial DNA were undertaken. It was shown that non-chromosomal *petite* mutants had incurred either loss or extensive damage to their mitochondrial DNAs. As a general rule neutral *petites* had completely lost mitochondrial DNA whereas suppressive *petites* usually contained fairly normal amounts of mitochondrial DNA, but this was often of an unusual buoyant density implying a gross alteration in mitochondrial DNA base composition. It has now been established by a number of studies that the usual situation in these *petites* is that the majority of the mitochondrial genome has been deleted and that the sequence of the remaining fraction of the genome has been repeated to such an extent that the total mitochondrial DNA level is approximately normal. *Petites* lacking in mitochondrial DNA are referred to as ρ^0 *petites*, those containing aberrant DNA as ρ^- while *grandes* are referred to as ρ^+. As we shall see in section 5.5 mitochondrial DNA codes for mitochondrial ribosomal RNA and so loss of one of the ribosomal RNA genes which invariably occurs in *petites*, precludes the possibility of any mitochondrial protein synthesis using messenger RNA derived from either nuclear or mitochondrial DNA. It is fortuitous that this inability is not lethal in bakers' yeast as it is in most other eukaryotes (including the majority of yeast species) as the *petite* mutation has contributed greatly to our knowledge of the function of the mitochondrial genome.

In addition to the *petite* a number of other mutations in bakers' yeast have been shown to affect the assembly of mitochondria. There are a range of mutants (*pet* mutants) which show normal chromosomal inheritance and which affect a variety of mitochondrial characteristics. Such mutants may be for a structural gene for a polypeptide component of one of the cytochromes for example. In fact, mutants lacking almost any cytochrome or any combination of cytochromes have been isolated. Other mutants with a chromosomal pattern of inheritance have been shown to be defective in the ability to synthesise ubiquinone, an iron-sulphur protein of Complex II, and the adenine nucleotide transporter.

Another series of mutants leading to defects in mitochondrial function in bakers' yeast are the *mit* mutants which are considered to be located on mitochondrial DNA. These mutations may lead to loss of cytochrome *b* (COB mutants), cytochrome aa_3 (OXI mutants), cytochromes *b* and aa_3 (BOX mutants) or ATP synthetase activity (PHO mutants). These mutants are established as due to lesions in mitochondrial DNA by a number of criteria:

1. The mutation does not segregate at meiosis in the manner expected of a nuclear gene.
2. Mitotic recombination occurs between these mutants and others known to be located on mitochondrial DNA (nuclear genes do not exhibit significant levels of mitotic recombination).
3. If mitochondrial DNA is eliminated by treatment with ethidium bromide to give *petite* mutants, the mutant characteristic is lost.

There is a further category of mitochondrial DNA mutations which can be recognised as such on the basis of these criteria. This category compromises mutants resistant to certain inhibitors of mitochondrial protein synthesis or ATP synthesis. Mitochondrial gene mutations lead to resistance to the protein synthesis inhibitors erythromycin, chloramphenicol, paromomycin and spiramycin and the ATP synthesis inhibitors oligomycin, venturicidin and triethyltin. The biochemical basis for these mutations will be discussed in section 5.5.

5.5   Mitochondrial assembly

The study of mitochondrial assembly is still at a fairly early stage and up to now the main questions which have stimulated interest are:

1. Which mitochondrial proteins are synthesised on mitochondrial ribosomes and which on cytoribosomes?
2. How is the integration of these components into the mitochondrial structure controlled?
3. What gene products are specified by mitochondrial DNA?
4. Why is it necessary to have decentralised mitochondrial DNA and protein synthesising systems?

The answers to all of these problems seem to be closely related. For example, there are four basic possible interactions of the two genetical systems (mitochondrial and nuclear DNAs) and two protein synthesising systems (mitoribosomal and cytoribosomal) involved in coding and synthesis of mitochondrial proteins. Mitochondrial DNA-derived messenger RNA might be translated on mitoribosomes or cytoribosomes as also might nuclear DNA-derived messenger RNA. Two of these four possibilities would demand that the mRNA would be transported across the inner membrane of the mitochondria. As we have seen this membrane is extremely impermeable even to fairly small molecules unless a special transporting system is available. At present there is no evidence that such carriers exist capable of selecting particular messenger RNAs for transport. In fact there is no really reliable evidence that any RNA molecules can traverse the inner membrane and it seems likely that any mitochondrial mRNA is translated on mitoribosomes and any nuclear-derived mRNA on cytoribosomes. If we make this assumption, and current evidence suggests that it is a valid one, it is possible to make some initial observations on the origin of mitochondrial proteins. We have already shown (sect. 5.3) that animal mitochondrial DNA is, at the most, capable of specifying about thirty small proteins (and yeast at most 150 proteins although in reality somewhat less than this). This number of proteins falls far short of the total mitochondrial protein complement and suggests that mitochondria can only be partially autonomous in terms of the proteins coded and synthesised within the organelle. What is more the coding potential for mitochondrial DNA for proteins is even less than suggested above because it is now firmly established that mitochondrial ribosomal RNAs and transfer RNAs are coded by the organelle genome. Both the large and small subunit rRNAs from a number of species have been shown to hybridise readily with mitochondrial DNA and only to a negligible extent with nuclear DNA. This is clear evidence for the complementarity of base sequences of mitochondrial rRNAs and mitochondrial DNA. Similar observations suggest that mitochondrial transfer RNAs are also coded by mitochondrial DNA. In species where considerable investigation of mitochondrial tRNAs has taken place 15–25 mitochondrial tRNAs have been identified. In some cases iso-accepting tRNAs for a single amino acid have been found. At one stage, a rather low number of tRNAs which could be found in Hela cell mitochondria led to a suggestion that, in this particular human cell line at least, mitochondrial proteins might not contain the usual range of amino acids. A more likely explanation would seem to be that further tRNAs remain to be discovered.

Petite mutants of yeast provide a very convenient way of studying what proteins are synthesised in mitochondria, and presumably coded by mitochondrial DNA. As we have seen, *petite* mutants usually retain only a small segment of the mitochondrial genome. Invariably this is insufficient to code for all the rRNAs and tRNAs essential for mitochondrial protein synthesis. Consequently any mitochondrial proteins found in petite mutants must be coded by nuclear DNA and synthesised on cytoribosomes. *Petite* mutants have been shown to contain a fairly normal complement of outer membrane and matrix enzymes, but have severe lesions in the mitochondrial ATP synthesising system. However, a number of electron carriers may be found in *petite* mutants. Succinate and NADH dehydrogenases are present but may be reduced either in amount or activity. Cytochromes c and c_1 are also present as are the F_1-ATPase and the OSCP (oligomycin-sensitivity conferring protein). It follows that all of the above must be coded and synthesised extramitochondrially. The problem remains of how these components are delivered to the inner membrane or matrix. Butow has observed what appear to be cytoribosomes associated with the outer membrane of yeast mitochondria at positions where the outer and inner membranes appear to come close together. It is tempting to suggest that these ribosomes are synthesising inner membrane or matrix components. The components which have been shown to be absent from the inner membrane in *petite* mutants are cytochromes b (i.e. $b562$ and $b566$) and cytochrome aa_3 and the polypeptides of the membrane sector of ATPase essential for its oligomycin-sensitivity. The absence of these components does not of itself prove that they are synthesised and coded intramitochondrially. An alternative explanation would be that they depend for their incorporation into the mitochondria on the previous incorporation of some other mitochondrially synthesised protein.

A more direct approach to finding out about the coding and protein

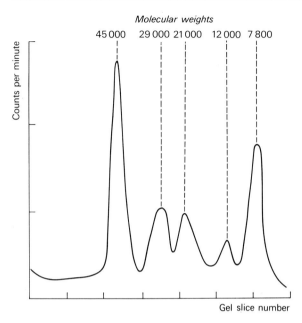

Fig. 5.7   Mitochondrial protein synthesis in yeast. Yeasts were grown in medium containing cycloheximide and [³H]-leucine. A mitochondrial preparation was made and dissolved in 1 per cent sodium dodecylsulphate (SDS) and analysed on an SDS-containing polyacrylamide gel. The gels were sliced, slices dissolved in hydrogen peroxide and counted in a scintillation counter.

synthesis properties of mitochondria is to investigate what proteins are synthesised either in isolated mitochondria or in intact cells whose cyto-ribosomal protein synthesis is inhibited. The use of isolated mitochondria has one important drawback. The rate of incorporation of radioactively labelled amino acids into mitochondria is very slow and may be limited by the lack of some unknown cytosolic factor. This low rate of incorporation also means that it is necessary to take precautions to maintain sterility to eliminate any contribution by bacterial contamination. This type of study showed that uptake of label by isolated mitochondria occurs solely into the inner membrane, a situation which was confirmed by later studies *in vivo*.

Once again yeasts have provided a very useful tool for studies *in vivo*. However, the structures of mitochondrially synthesised inner membrane components of yeast are very similar to those found in higher organisms and it seems likely that observations made on yeast may have more general application. Much of what we know about the individual proteins syn-

129

thesised in yeast mitochondria comes from a penetrating series of experiments carried out by Tzagaloff and his co-workers. The basic technique was to incubate yeast with radioactively labelled leucine in the presence of cycloheximide which completely blocks cytoribosomal, but permits mitoribosomal protein synthesis. In initial experiments, submitochondrial particles were prepared from the labelled cells and then dissolved in sodium dodecyl sulphate (SDS) and resolved by polyacrylamide gel electrophoresis. The gels were sliced and resulting slices dissolved and counted for radioactivity in a scintillation counter. Five major bands of labelling were observed varying in molecular weights from 45 000 to 8 000 (Fig. 5.7). More recent sophistications to this technique, involving more effective labelling conditions and the use of polyacrylamide gradient slab gels examined by autoradiography, have shown that there may be as many as twenty different proteins synthesised in yeast mitochondria. An indication of what the main products were came from experiments in which individual enzyme complexes of the oxidative phosphorylation system were prepared from yeast incubated in the presence of radioactively labelled leucine and either chloramphenicol or cycloheximide and resolved on SDS polyacrylamide gels. Two major enzyme complexes have been studied in detail in this way — the oligomycin-sensitive ATPase and cytochrome oxidase complexes. The oligomycin-sensitive ATPase from yeast was found initially to be composed of ten types of subunit: five F_1-ATPase subunits ($F_1.1$ to $F_1.5$), four membrane sector subunits (MS.1 to MS.4) and the OSCP (see also Fig. 3.6). The mitochondrial inhibitor from yeast has only very recently been observed. The five F_1 subunits and the OSCP were found to be labelled in the presence of chloramphenicol which suggests that they are synthesised on cytoribosomes, a conclusion which supports the observation of their presence in *petite* mutants. The four subunits of the membrane sector are labelled in the presence of cycloheximide but not chloramphenicol and are thus synthesised by mitoribosomes. These subunits, which are highly hydrophobic, are thought to be the base piece of the tripartite ATPase particles of the inner membrane. Subunit MS.4, for example, contains about 78 per cent amino acids with hydrophobic side chains. Recent work has suggested that the mitochondrial inhibitor may be absent from *petite* mutants and this may possibly be a product of mitochondrial protein synthesis.

Using similar techniques it has been shown that cytochrome oxidase of yeast is composed of seven subunits (Table 5.3) of which the three largest are synthesised within the mitochondria. These three subunits are also very hydrophobic.

The only other clearly established product of mitochondrial protein synthesis is one subunit of the cytochrome bc_1 complex, probably the apoprotein of the cytochromes b. As far as we are aware at present, the rest of the mitochondrial electron transport system can be assembled without the intervention of mitochondrial protein synthesis. What then are

Table 5.3 Subunit composition of cytochrome oxidase of yeast

Subunit	Molecular weight	Synthesis
1	40 000	Mitochondrial
2a	27 300	Mitochondrial
2b	25 000	Mitochondrial
3a	13 800	Extramitochondrial
3b	13 000	Extramitochondrial
4	10 200	Extramitochondrial
5	9 500	Extramitochondrial

the other products of mitochondrial protein synthesis which have recently been observed in high resolution techniques? At present it is not known; they could possibly be artefacts but it is also speculated that they may be involved as controlling elements in mitochondrial macromolecular synthesis. Growth of yeast, either in the presence of high glucose (or sucrose or fructose) or under anaerobic conditions shuts off the synthesis of many of the enzymes of the oxidative phosphorylation system including all of the cytochromes. It is possible that minor products of mitochondrial protein synthesis may be involved in these control processes.

The hydrophobicity of the known mitochondrially-synthesised proteins may provide an indication of the necessity for mitochondrial protein synthesis. Their lack of solubility in an aqueous environment may require that they be synthesised very close to their site of operation, in the inner membrane. If, as we speculated earlier, the mitochondrial membrane is a barrier to nucleic acid transport, mitochondrial protein synthesis would require intramitochondrial assembly of ribosomes and tRNAs and the presence and replication of DNA.

As well as the protein components synthesised by the two ribosomal systems of the cells the inner membrane of mitochondria contains lipids and carbohydrates (as components of glycoproteins) as integral to its structure and essential for its function. Little progress has been made so far in determining the interactions of these various components in mitochondrial assembly. It is known, however, that the overall protein composition of the mitochondria is little altered in *petite* mutants which are completely lacking in mitochondrial protein synthesis. Even those proteins (NADH dehydrogenase, F_1ATPase, cytochrome *c*) which are intimately associated with mitochondrially synthesised proteins are still present in *petite* mutants at relatively normal levels, implying a lack of control by the mitochondrial genome and its products, on cytoribosomal synthesis of mitochondrial components. The opposite does not appear to be the case, however, as the rate of mitochondrial protein synthesis seem to be limited by the availability of extramitochondrial products. Figure 5.8(*a*) and 5.8(*b*) summarise the current status of the various interactions involved in the assembly of the ATP synthesising system.

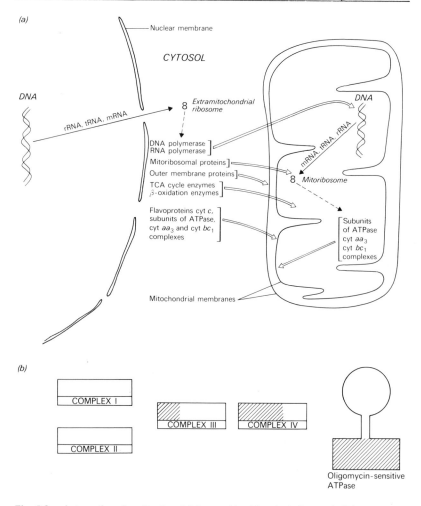

Fig. 5.8 Interactions in mitochondrial assembly. The shaded areas in (*b*) represent approximately the proportions of proteins in individual mitochondrial inner membrane complexes which are synthesised on mitochondrial ribosomes.

5.6 Mitochondrial genetics

The presence of a genome within the mitochondria has prompted attempts to construct genetic maps of mitochondrial DNA for different species. As with other areas of study of mitochondrial biogenesis, bakers' yeast (*Saccharomyces cerevisiae*) has figured prominently in these investigations.

A number of techniques have been used in the study of yeast mitochondrial genetics. The most direct technique is to cross strains of opposite mating type and with contrasting mitochondrial gene markers and to analyse diploids derived from zygotes for recombination of the mitochondrial genes. If a chloramphenicol-resistant / erythromycin-sensitive ($C^R E^S$) strain is crossed with a chloramphenicol-sensitive / erythromycin-resistant ($C^S E^R$) strain a certain proportion of the diploids produced will be $C^R E^R$ or $C^S E^S$. Generally the proportion of recombinants in the diploid progeny may be taken as an indication of the linkage of the genes — the higher the proportion of recombinants, the greater the map distance between the genes. The interpretation of such results is complicated by the observation that, in certain crosses, following a period when the zygote gives rise to diploid buds with any of the possible gene combinations it subsequently produces diploids each having the same combination as one of the parents. This may be related to another phenomenon known as polarity. The polarity of transmission of certain genes into the diploids appears to be determined by a gene locus (the ω locus) on the mitochondrial DNA. This is present either as an ω^+ or an ω^- allele. In $\omega^+ \times \omega^-$ crosses (heterosexual crosses) alleles closely linked to the ω^+ are preferentially transmitted. Thus, in such crosses, equal numbers of either the two parental or two recombinant gene combinations are not produced. For example, in the cross $\omega^+ C^R E^S \times \omega^- C^S E^R$ the parental combination $C^R E^S$ is produced in considerably greater numbers than the other parental combination $C^S E^R$ and there are more $C^R E^R$ recombinants than $C^S E^S$ because of the close linkage of the ω^+ and C^R alleles. In $\omega^+ \times \omega^+$ or $\omega^- \times \omega^-$ crosses (homosexual crosses) such polarity does not exist. The reason for this behaviour is not yet established but it is a factor which must be considered when setting up crosses.

Another way in which linkage relationships of yeast mitochondrial genes may be studied is to look for the co-retention of genes in *petite* mutants. As we saw in section 5.4 the ρ^- *petites* retain and amplify a small randomly selected segment of the genome. If two mutations are both retained in the *petite* the implication is that they are located in the same segment of mitochondrial DNA and are thus closely linked. To test for the presence of mutant genes in the *petite* mitochondrial DNA, it would be necessary to cross the *petite* with a wild-type strain and examine the mutant progeny for the mutant characteristic. This is necessary because the lack of a mitochondrial protein-synthesising system in the ρ^- strain prevents expression of the mutant genes.

A recent technique which has provided some information on the localisation of the mitochondrial genes has been to establish a physical map of yeast mitochondrial DNA by cleaving the molecule into a discrete number of fragments by treatment with restriction endonucleases which split the molecules at particular sites. The fragments can be sized either by electron microscopy or by agarose gel electophoresis and the linear

133

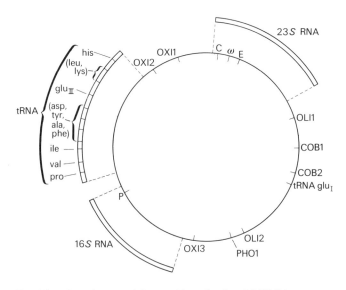

Fig. 5.9 Genetic map of *S. cerevisiae* mitochondrial DNA.
C = chloramphenicol resistance; E = erythromycin resistance; P = paromomycin resistance; OLI1 and OLI2 = oligomycin resistance; ω = polarity locus; COB = cytochrome *b*; OXI1, OXI2 and OXI3 = cytochrome oxidase; PHO = ATP synthetase.

arrangement of the fragments in the parent molecule can be determined by sizing the products of incomplete digestion by the endonuclease. By looking at the hybridisation of ribosomal RNAs with the different fragments it has been possible to locate the ribosomal RNA genes on the mitochondrial DNA. Similarly, by looking at hybridisation between ρ⁻ mitochondrial DNA retaining known genes and the endonuclease fragments, it has been possible to obtain a rough idea of the localisation of the genes. Figure 5.9 shows the current status of the mitochondrial DNA gene map for *Saccharomyces cerevisiae*. It is interesting to note that the erythromycin-resistance gene falls within the 23S ribosomal RNA gene. Thus, unlike the situation in bacteria where resistance to this antibiotic is associated with an altered ribosomal protein, in yeast mitochondria resistance results from an altered rRNA.

Progress in mapping mitochondrial DNAs in other species is nowhere near so advanced as in yeast. However, non-chromosomal mutations equivalent to those found in yeast are being studied in *Neurospora*, *Aspergillus* and *Paramecium*. Furthermore, the relatively short contour length of animal mitochondrial DNAs make them very amenable to physical mapping of rRNAs and tRNAs by hybridisation with endonuclease fragments.

134

Despite the fact that mitochondrial DNAs were only fairly recently dis-covered, it seems likely that they will become the first eukaryotic genetic elements to be completely characterised.

5.7 Evolutionary origin of mitochondria

In this book we have examined the structure, function and assembly of the mitochondrion. We have built up a picture of an extremely complex organelle whose modes of operation and assembly are still incompletely understood. We feel that this final section is an appropriate place to ask the question, 'How did this organelle evolve?' As was observed at the very beginning of the book, living organisms are either prokaryotes and eukaryotes, and as far as we are aware, there are no intermediate forms. With respect to mitochondria the situation is very simple — prokaryotes lack mitochondria, eukaryotes contain them. Thus one of the steps in the evolution of eukaryotic from prokaryotic cells must have been the transfer of the apparatus for synthesis of ATP coupled to electron transport, from the limiting membrane of the prokaryote to an intracellular site in discrete vesicular particles which we know as mitochondria. At the same time the mitochondria have somehow acquired a distinct genome and protein-synthesising system. The presence of the latter elements in mitochondria has led to the suggestion that mitochondria evolved from the invasion of a relatively large non-aerobic prokaryote by a smaller aerobic bacterium. As photosynthetic organisms caused the oxygen concentration of the atmos-phere to rise it is possible that the host-invader relationship of the two species would give way to a symbiotic association wherein oxidatively produced ATP of the aerobic bacterium would be given up to the host in exchange for the relatively congenial surroundings of the host cytosol. Now, it is well known that endoparasites, whose way of life is not usually very demanding tend to degenerate and lose certain functions not essential to their existence. Thus the aerobic endosymbiont, in this case, might lose all of those genes and processes not essential to its relationship with the host cell. In other words it might specialise in ATP synthesis. Broadly, this is what has been termed the endosymbiotic theory for the origin of mitochondria — a cause which has been ably championed by Margulis. She has drawn attention to considerable similarities between mitochondria and bacteria. They are of similar size and both have a circular genome. The protein synthesising systems of both mitochondria and bacteria are sensitive to erythromycin and chloramphenicol and insensitive to cycloheximide and emetine whilst cytoribosomal protein synthesis is insensitive to erythromycin and chlor-amphenicol and usually sensitive to cycloheximide and emetine. In mito-chondria and bacteria protein synthesis is initiated by the binding of N-formylmethionyl-tRNA to the small ribosomal subunit. In cytoribosomal protein synthesis the initiating aminoacyl-tRNA is methionyl-tRNA.

In section 5.5 we saw that the great bulk of mitochondrial proteins are synthesised by cytoribosomes translating nuclear-derived messenger RNA. The endosymbiotic theory would require that a large number of the endosymbiont's genes must have been transferred to the host nucleus. It is very difficult to see why some of the structural genes for the oligomycin-sensitive ATPase, for example, should be transferred and not others.

An alternative hypothesis for mitochondrial origin is that an invagination of the prokaryotic limiting membrane may have become nipped off to form a vesicle. If on its formation this vesicle had happened to enclose a suitable piece of DNA which had been excised from the main DNA a protomitochondrion could have been formed. Such excised pieces of DNA (episomes or plasmids) are now frequently found in bacterial strains. This is the basis of the episome theory for mitochondrial origin whose main proponent has been Mahler. The similarities between mitochondria and bacteria can just as adequately be explained on the basis of the episome theory. Previous to the time when mitochondria first appeared all organisms were prokaryotes. Consequently the DNA and protein-synthesising system isolated within the protomitochondrion, no matter what its origin, must have been of a prokaryotic nature. What has happened in the course of evolution is that the nuclear and cytosolic systems have undergone evolution to their present day characteristics while the intramitochondrial prokaryotic features have been conserved.

We do not really know how mitochondria evolved, but the problem has generated some stimulating discussion. What can be said with some confidence is that the evolution of mitochondria was one of the crucial factors which has permitted the development of the incredible variety and complexity that we see in present day eukaryotes.

Suggested further reading

Books
LLOYD, D., (1974) *The Mitochondria of Micro-organisms.* Academic Press, London and New York.
MARGULIS, L. (1970) *Origin of Eukaryotic Cells.* Yale University Press.
ROODYN, D. B. and WILKIE, D. (1968) *The Biogenesis of Mitochondria.* Methuen, London.
SAGER, R. (1972) *Cytoplasmic Genes and Organelles.* Academic Press, London and New York.

General reviews
ASHWELL, M. and WORK, T. S. (1970) The biogenesis of mitochondria, *Ann. Rev. Biochem.,* **39**, 251—70.
LINNANE, A. W., HASLAM, J. M., LUKINS, H. B. and NAGLEY, P. (1972) The biogenesis of mitochondria in microorganisms, *Ann. Rev. Microbiol.,* **26**, 163—98.
MAHLER, H. R. (1973) *Mitochondria: Molecular Biology, Genetics and Development. An Addison-Wesley Module in Biology,* No. 1.
RABINOWITZ, M. and SWIFT, H. (1970) Mitochondrial nucleic acids and their relation to the biogenesis of mitochondria, *Physiol. Rev.,* **60**, 376—427.

Cellular origin or mitochondria

LARSEN, W. J. (1970) Genesis of mitochondria in insect fat body, *J. Cell. Biol.*, **47**, 373—83.

LUCK, D. J. L. (1963) Formation of mitochondria in *Neurospora crassa*, *J. Cell. Biol.*, **16**, 483—99.

Mitochondrial nucleic acids and protein synthesis

ALONI, Y. and ATTARDI, G. (1972) Expression of the mitochondrial genome in Hela cells. XI. Isolation and characterization of transcription complexes of mitochondrial DNA, *J. Mol. Biol.*, **70**, 363—73.

BORST, P. (1972) Mitochondrial nucleic acids, *Ann. Rev. Biochem.*, **41**, 333—76.

ECCLESHALL, T. R. and CRIDDLE, R. S. (1974) The DNA-dependent RNA polymerases from yeast mitochondria, in *Biogenesis of Mitochondria*, pp. 31—46, eds. A. M. Kroon and C. Saccone. Academic Press, London and New York.

EHRLICH, S. D., THIERRY, J.-P. and BERNARDI, G. (1972) The mitochondrial genome of wild-type yeast cells. III. The pyrimidine tracts of mitochondrial DNA, *J. Mol. Biol.*, **65**, 207—12.

NAGLEY, P. and LINNANE, A. W. (1970) Mitochondrial DNA deficient *petite* mutants of yeast, *Biochem. Biophys. Res. Commun.*, **39**, 989—96.

NASS, M. M. K. and NASS, S. (1963) Intramitochondrial fibers with DNA characteristics. I. Fixation and electron staining reactions, *J. Cell. Biol.*, **19**, 613—29.

TEWARI, K. K., JAYARMAN, J. and MAHLER, H. R. (1965) Separation and characterization of mitochondrial DNA from yeast, *Biochem. Biophys. Res. Commun.*, **21**, 141—7.

Mitochondrial mutants

EPHRUSSI, B. (1952) The interplay of heredity and environment in the synthesis of respiratory enzymes in yeast. *Harvey Lectures, Series XLVI*, pp. 45—67.

EPHRUSSI, B., MARGERIE-HOTTINGUER, H. and ROMAN, H. (1955). Suppressiveness: a new factor in the genetic determinism of the synthesis of respiratory enzymes in yeast, *Proc. Natl. Acad. Sci.*, **41**, 1065—71.

FOURY, F. and TZAGALOFF, A. (1976) Localization on mitochondrial DNA of mutations leading to a loss of rutamycin-sensitive adenosine triphosphatase, *Eur. J. Biochem.*, **68**, 113—19.

SLONIMSKI, P. P. and TZAGALOFF, A. (1976) Localization in yeast mitochondrial DNA of mutations expressed in a deficiency of cytochrome oxidase and/or coenzyme QH_2 — cytochrome c reductase, *Eur. J. Biochem.*, **61**, 27—41.

Mitochondrial assembly

KELLEMS, R. E. and BUTOW, R. A. (1972) Cytoplasmic 80 *S* ribosomes associated with yeast mitochondria. I. Evidence for ribosome-binding sites on yeast mitochondria, *J. Biol. Chem.*, **247**, 8043—50.

TZAGALOFF, A., RUBIN, M. S. and SIERRA, M. F. (1973) Biosynthesis of mitochondrial enzymes, *Biochim. Biophys. Acta*, **301**, 71—104.

Mitochondrial genetics

COEN, D., DEUTSCH, J., NETTER, P., PETROCHILO, E. and SLONIMSKI, P. P. (1970) Mitochondrial genetics. I. Methodology and phenomenology, in *Control of Organelle Development*, ed. P. L. Miller. SEB symposium No. 24, pp. 449—96, Cambridge University Press.

GILLHAM, N. W. (1974) Genetic analysis of the chloroplast and mitochondrial genomes, *Ann. Rev. Genet.*, **8**, 347—91.

WILKIE, D. and THOMAS, D. Y. (1973) Mitochondrial genetic analysis by zygote cell lineages in *Saccharomyces cerevisiae*, *Genetics*, **73**, 367—77.

Evolution of mitochondria
MARGULIS, L. (1975) Symbiotic theory for the origin of eukaryotic organelles: criteria for proof, in *Symbiosis*, eds. D. H. Jennings and D. L. Lee. SEB symposium No. 29, pp. 21–38. Cambridge University Press.
RAFF, R. A. and MAHLER, H. R. (1975) The symbiont that never was: an enquiry into the evolutionary origin of the mitochondrion, in *Symbiosis*, eds. D. H. Jennings and D. L. Lee. SEB symposium No. 29, pp. 41–92. Cambridge University Press.

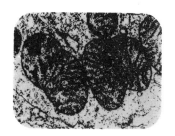

Index

Acetate, 23, 24
Acetoacetate, 45, 100
Acetoacetyl CoA, 45
Acetyl CoA, 23–6, 28–31, 33, 34, 36, 40–3, 92
Acetyl CoA: acetyl transferase, 39, 40, 45
N-acetyl glutamate, 37
Aconitase, 24
cis-Aconitate, 24
Acriflavine, 122
Acyl CoA dehydrogenase, 39, 40
Adenine nucleotide transporter, 96, 98, 99, 105, 106, 110
Adenosine diphosphate (ADP), 36, 42, 43
Adenosine monophosphate (AMP), 31
Adenosine monophosphate deaminase, 31
Adenosine triphosphate (ATP), 17–19, 23, 24, 40, 41
  phosphate donor potential, 19
  structure, 18
  synthesis, 47
Adenosine triphosphatase (ATPase)
  assay, 82
  calcium dependent, 82
  mitochondrial, 47, 66, 74, 80–7, 92, 102, 110, 126
  muscle, 18
  Na:K dependent, 18, 82
  synthesis of ATP, 47
Adenylate kinase, 16, 20

Adenylosuccinate, 31
Adenylosuccinate lyase, 31
Adenylosuccinate synthetase, 31
Adipose tissue, 30, 38
ADP:O ratio (ADP:2e⁻ ratio), 70
Adrenaline, 38
Alanine, 25, 29
Alanine amino transferase (GPT), 28, 29
Amino acid synthesis, 26, 28
Amino acyl tRNA synthetase, 20, 121
Ammonia, 28, 29, 31, 36–8
Amytal, 61, 66
Anaplerotic reactions, 22, 29, 30, 31, 42
Anilinonaphthalene sulphonic acid (ANS), 80
Antibodies, 63, 64, 82
Antimycin A, 61, 65, 66, 69, 70, 92, 95
  structure, 66
Arginine, 37, 38
Arginase, 37
Arginino succinate, 37
Arginino succinate lyase, 37
Arginino succinate synthetase, 37
Ascorbate/TMPD, 61, 69, 70
Aspartate, 25, 29, 31, 36–8
  transporter, 97, 101, 103, 104
Aspartate amino transferase (GOT), 28 29, 103, 106
Aspergillus, 134
ATP-citrate lyase, 28
Atractyloside, 96, 99
Aurovertin, 86, 87

Azide, 61, 64, 66

Bacteriorhodopsin, 80, 81
Biological work, 17–19
Biotin, 29
Bongkrekic acid, 96, 99
Borst cycle, 103
Brain, 29
n-Butyl malonate, 96, 99, 101

Caesium chloride gradient centrifugation, 115, 116
Calcium, 8, 9, 36, 42, 86, 97, 108, 110
Calpodes ethlius, 113, 114
Carbamoyl phosphate, 37
Carbamoyl phosphate synthetase, 19, 37
Carbon monoxide, 54, 59, 61, 66
Carbonyl cyanide-p-trifluoromethoxy-phenylhydrazone (FCCP), 67, 69, 76
    structure, 67
Carboxyatractyloside, 96, 99
Cardiolipin, 16, 91, 110
Carnitine acyl transferase, 38, 39, 106
    location, 38, 39
Carriers, 17, 44, 91
    see also Transporting systems
Chemical coupling theory, 71–3, 80
Chemiosmotic coupling theory, 73–80
    vectorial Bohr theory, 78, 79
Chloramphenicol, 120, 127, 130, 133–5
Choline, 113–15
Chromulina pusilla, 3, 113
Cilia, 17
Citrate, 23–8, 30, 31
Citrate lyase, 28, 31
Citrate synthase, 24, 28, 31, 41
    control, 42
Citric acid cycle, 22
    see also Tricarboxylic acid cycle
Citrulline, 37, 92
    transporter, 97, 105
Coenzyme A, 24, 30–4, 36
    structure, 32
Conformational coupling theory, 79, 80
    conformational changes, 80
    electromechanochemical model, 79, 80
Copper, 59
Coupling factors, 82
Cristae, 5, 9, 63
Crossover point, 65, 66
Cyanide, 54, 59, 61, 66, 69, 70, 88
α-cyano-4-hydroxycinnamate, 96, 100
Cycloheximide, 121, 130, 135
Cytidine triphosphate (CTP), 20

Cytochrome $a.a_3$, 54–6, 59, 61, 62, 64, 65, 73, 74, 126, 128
    absorption peaks, 57
    location, 64
    redox potential, 60, 73
Cytochrome b, 56, 59, 61, 62, 64, 65, 73, 74, 126, 130
    location, 64
Cytochrome $b.c_1$ complex (Complex III), 51, 61, 62, 66, 70, 88, 132
    biosynthesis, 130, 132
Cytochrome b-562 (b_K), 55, 59, 77, 78, 128
    absorption peaks, 57
    redox potential, 60
Cytochrome b-566 (b_T), 55, 59, 73, 77, 78, 128
    absorption peaks, 57
    redox potential, 60, 73
Cytochrome c, 55, 56, 59, 61, 62, 64, 65, 73, 128, 131, 132
    absorption spectrum, 57
    biosynthesis, 128, 132
    location, 55, 64
    redox potential, 60
Cytochrome c_1, 55, 59, 64, 78, 128
    absorption peaks, 57
    biosynthesis, 128
    location, 64
    redox potential, 60
Cytochrome oxidase complex (Complex IV), 59, 61, 62, 64, 66, 70, 78, 132, 134
    biosynthesis, 130–2
    composition, 131
Cytochrome P450, 45, 54, 55, 88
Cytochromes, 26, 54–9, 126
    absorption spectra, 56, 57
    location, 55
    oxidation-reduction, 56, 57

Diaminobenzene sulphonic acid (DABS), 64, 65
Dicarboxylate transporter, 27, 30, 31, 93, 96, 100
Dicyclohexylcarbodiimide (DCCD), 67, 86
    structure, 67, 83
Difference spectrum 56–8
Digitonin, 13, 63, 65
Dihydrolipoyl dehydrogenase, 50
    see also lipoamide dehydrogenase
2.4-Dinitrophenol (DNP), 67, 74
    structure, 67
DNA polymerase, 120

Drosophila melanogaster, 117
Dual wavelength spectrophotometry, 59

Electron spin resonance (ESR), 50, 51, 59
Electron transferring flavoprotein (ETF), 40, 50, 61
  location, 40
Electron transport system, 22, 47–67, 76
  functional organisation, 61, 62, 65, 75, 77, 78
  inhibition, 63, 65
  initiation, 22, 47, 49
  location of carriers, 13, 47
  redox state of carriers, 61, 62
  reversibility, 87
  structural organisation, 61–5, 73
Emetine, 121, 135
Endoplasmic reticulum, 45
Endosymbiosis, 3, 135, 136
Enoyl CoA, 39, 40
Enoyl CoA hydratase, 39, 40
Enoyl CoA isomerase, 40
Episome theory, 136
Erythromycin, 121, 127, 133–5
Escherichia coli, 34
Ethidium bromide, 122
2-Ethyl citrate, 100–1
N-ethyl maleimide (NEM), 96, 97, 99–101
Euglena gracilis, 117
Exchange reactions, 83, 84
  ADP-ATP exchange, 83, 86
  Pi-ATP exchange, 83, 86
  Pi-H_2O exchange, 83

F_1-ATPase, 64, 65, 81, 82, 84–6, 106, 128, 131
  biosynthesis, 130
  composition, 86, 87
  properties, 86, 87
Fatty acid oxidation (β-oxidation), 17, 22, 25, 38, 93, 106
  control, 43
  location, 16, 25
Fatty acid synthesis, 13, 27, 30, 31, 92
Fatty acid synthetase complex, 27, 92
Fatty acyl CoA, 38–40, 60, 92
  transporting system, 97, 106, 107
Fatty acyl thiokinase, 38, 39
  location, 38, 39
Fenestrations, 7
Ferricyanide, 65
Ferrochelatase, 45

Flagellae, 17
Flavin adenine dinucleotide (FAD), 22–4, 32–4, 39–41, 50, 60, 74, 75
  oxidation-reduction, 50, 51
  structure, 51
Flavin mononucleotide (FMN), 50–2, 73–5
  oxidation-reduction, 50, 51
Flavoproteins, 45, 50, 61, 62
  absorption spectrum, 52
N-formyl methionyl tRNA, 135
Fumarase, 24
Fumarate, 24–6, 30, 31, 47, 50, 60

Glisoxepide, 97, 101
Glucagon, 100
Gluconeogenesis, 26, 27, 29, 96, 102
Glutamate, 25, 26, 28, 29, 36, 38, 61, 69, 93
  transporting system, 94, 95, 97, 101, 104, 110
Glutamate dehydrogenase, 28, 31, 36, 38, 50, 94
Glutamic oxaloacetic transaminase (GOT), 29
Glutamic pyruvic transaminase (GPT), 29
Glutamine, 105, 106
  transporting system, 97, 98, 105, 106
Glutamine synthetase, 19
L-glycerol-3-phosphate (α-glycerophosphate), 61, 103, 104
L-glycerol-3-phosphate dehydrogenase (α-glycerophosphate dehydrogenase), 50, 103–5
Glycolysis, 3, 25, 29, 31, 71, 92, 102
  reversal, 26
Glyoxylate, 26, 30
  bypass, 30, 41
Glyoxysomes, 30
Gramicidin, 107, 108
Guanosine diphosphate (GDP), 23, 24, 27, 31, 86
Guanosine triphosphate (GTP), 19, 23, 24, 27, 31

Haem, 54
  biosynthesis, 45
Haem A, 55, 56
Haem C, 55, 56
Halobacterium halobium, 80, 81
  purple membrane, 80, 81
Hexokinase, 70

3-hydroxyacyl CoA, 39, 40
3-hydroxyacyl CoA dehydrogenase, 39, 40, 50
3-hydroxybutyrate, 45, 76
3-hydroxybutyrate dehydrogenase, 45, 50
β-hydroxy-β-methyl glutaryl CoA, 45
  lyase, 45
  synthetase, 45

Inner membrane, 8, 9, 113, 115, 120, 128, 129
  'C' side, 63−5, 73, 75−7, 82
  density, 13
  enzyme content, 13, 32
  isolation, 13
  morphology, 5, 7
  'M' side, 63−5, 73, 75, 76, 78, 81, 82
  permeability, 16, 27, 127
  presence of electron carriers, 13, 49
Intermembrane space, 4, 8, 13, 16, 20, 45
Inosine monophosphate (IMP), 31
Inosine triphosphate (ITP), 86
Intracristal space, 8
Ionophores, 107
Iron sulphur centres, 51
  oxidation-reduction, 51
  structure, 53
Iron sulphur proteins, 32, 45, 50, 51, 73, 126
  oxidation-reduction, 51
Isocitrate, 24-6, 30, 60, 61, 93
  transporting system, 94, 96
Isocitrate dehydrogenase (NAD-linked), 23, 24, 41, 94, 109
  control, 42, 43
Isocitrate lyase, 26, 30

3-ketoacyl CoA, 39, 40
α-ketoglutarate, 23
  see also 2-oxoglutarate
Ketone bodies, 45, 46
Kluyveromyces lactis, 56, 58, 117
Krebs cycle, 22
  see also Tricarboxylic acid cycle
Kynurenine hydroxylase, 13

Lactoperoxidase, 64
Lanthanum ions, 97, 108−10
Lipoate acetyl transferase, 33−5
  lipoyllysyl side chain, 33−6
Lipoamide dehydrogenase, 33−6
Lipoic acid (LA) 32−5, 73
Luciferase, 82

Magnesium, 36
Malate, 24−8, 30, 31, 47, 48, 60, 61, 69, 70
Malate dehydrogenase, 23, 24, 26, 28, 31, 50, 60, 103
  cytosol, 27
  matrix, 27
  reaction, 48
  stereospecificity, 49
Malate synthase, 26, 30
Malic enzyme (Malate dehydrogenase decarboxylating), 30, 31
Membrane potential, 73, 75
Membrane sector, 64, 84, 128
  biosynthesis, 130
  composition, 86, 87
Mersalyl, 96, 99
Mesosomes, 1
Mit mutants, 126
Mitochondria
  adrenal cortex, 5, 12, 45, 88
  assembly, 112−36
  bird flight muscle, 5
  brown adipose tissue, 71
  cellular origin, 113−15
  distribution, 1, 3
  division, 113, 114
  Euglena, 16
  evolution, 1, 4, 135−6
  heart muscle, 5, 9, 15, 33
  insect flight muscle, 3−7, 8, 12, 14, 15, 56, 58, 64, 69
  isolation, 9−11, 17
  liver fluke, 5, 10
  plant, 5, 11
  rat-anterior-pituitary, 4, 6
  rat-liver, 3−7, 60
  shape, 4
  size, 4
  spermatozoa, 3
  structure, 3
  tape worm, 5
  yeast, 2, 3
Mitochondrial assembly, 127−32
Mitochondrial DNA, 17, 112, 113, 115−21, 125−8, 132−4
  coding capacity, 118, 120, 127, 128
  densities, 115−17
  electron microscopy, 115, 118−21
  genetic analysis, 132−5
  location, 16
  mutations in, 126, 127
  quantity, 118
  replication, 120, 121, 131
  sizes, 117, 118

synthesis, 16
transcription, 120
Mitochondrial genetics, 132–5
  DNA:RNA hybridisation, 134
  genetic map, 134
  linkage, 133
  petite deletion analysis, 133
  polarity, 133, 134
  recombination, 133
Mitochondrial inhibitor (MI), 86, 87,
  130
  biosynthesis, 130
Mitochondrial matrix, 4, 8, 13, 75
  enzyme content, 16, 128
Mitochondrial metabolism,
  regulation, 41
Mitochondrial mutants, 122–7
Mitochondrial protein synthesis, 17, 112,
  113, 127–33, 135
  control, 131
  elongation factors, 120
  inhibition, 121, 127
  initiation factors, 120
  role, 131
Mitochondrial ribosomes, 120, 127,
  130–2
  location, 9, 120
  size, 120
Mitochondrial RNA, 17, 120
  messenger RNA, 120, 121, 132
  ribosomal RNA, 120, 126–8, 132,
    134
  synthesis, 16
  transfer RNA, 121, 128, 131, 133,
    134
Mitochondrial swelling, 13, 17, 58
Monoamine oxidase, 13
Musca domestica, 117
Muscle contraction, 18, 108

NAD:NADP transhydrogenase, 13, 88
NADH dehydrogenase, 50, 51, 64–6, 73,
  75, 128, 131
NADH dehydrogenase complex
  (Complex I), 61, 62, 66, 70, 88,
  132
Neurospora crassa, 113, 134
Niacin, 48
Nicotinamide adenine dinucleotide
  (NAD), 22–4, 27, 28, 31–4, 36,
  38, 40, 41, 48, 49, 60–2, 65, 70,
  72, 74, 75, 93–5
  absorption spectrum, 49
  oxidation-reduction, 49
  redox potential, 60

structure, 48
  transfer of reducing equivalents,
    102–5
Nicotinamide adenine dinucleotide
  phosphate (NADP), 30, 31, 48, 88
Nigericin, 107, 108
Non-equilibrium reactions, 41
Non-haem iron, 32
  see also Iron sulphur proteins
Nucleoside diphosphokinase, 16, 20, 23

Oleic acid, 73
Oleyl lipoate, 73
Oleyl phosphate, 73
Oligomycin, 67, 69, 72, 76, 83, 86, 88,
  99, 127, 134
Oligomycin sensitivity conferring
  protein (OSCP), 64, 81, 82, 84,
    86, 87, 128
  biosynthesis, 130
Oligomycin-sensitive ATPase (ATP
  synthetase), 82, 84–7, 134, 136
  biosynthesis, 130, 132
  composition, 86, 130
  inhibition, 86
  preparation, 85
  properties, 85, 86
  structure, 84, 86, 87
Ornithine, 37, 92
  transporting system, 97, 105
Ornithine carbamoyl transferase
  (ornithine transcarbamylase), 37,
    43
Outer membrane
  biosynthesis, 132
  density, 13
  enzyme content, 13, 43, 128
  isolation, 13
  morphology, 5
  permeability, 16
Oxaloacetate, 23–31, 42, 47, 48, 60,
  101
Oxidative decarboxylation, 32
Oxidative phosphorylation, 47–88
  coupling, 70–81
  efficiency, 71
  importance, 20
  inhibitors, 67, 69, 72, 86
  measurement, 68–70
  reversibility, 83
  specificity, 20
  uncouplers, 67, 69, 72, 76, 82, 83, 88
2-oxoglutarate, 23–6, 28, 29, 32, 38, 60
  transporting system, 97, 100, 101,
    103

2-oxoglutarate decarboxylase, 36, 41
  control, 42
2-oxoglutarate dehydrogenase, 23
  complex, 24, 32, 36
Oxygen electrode, 68—70

Paramecium aurelia, 117, 134
Paromomycin, 127, 134
Pediculi, 8 & 16
Pet mutants, 126
Petite mutants, 3, 122—6, 128, 131
  genetics, 122—4
  induction, 122
  mitochondrial DNA, 125, 126, 128,
    133
  neutral, 124, 125
  nuclear (segregational), 123—5
  properties, 122, 128, 131
  ρ^0 petites, 126
  ρ^- petites, 126
  suppressive, 124, 125
Phenylketonuria, 100
Phenyl pyruvate, 100
2-phenyl succinate, 96, 101
pH gradient, 75, 81
Phosphate donor potential, 19, 71
Phosphate transporter, 77, 93, 96, 98,
  100, 102
Phosphoenolpyruvate, 26—8, 41, 101
Phosphoenolpyruvate carboxykinase, 27,
  41
Phosphofructokinase, 96, 102
3-Phosphoglyceraldehyde dehydrogenase,
  71, 72
Phosphoglycerate kinase, 71, 72
Phospholipids, 13, 20, 26, 87, 91, 109,
  110, 113
  vesicles, 81
Piericidin, 61, 66
Pisum sativum, 117
Polyacrylamide gel electrophoresis, 129,
  130
Polylysine, 65
P:O ratio (P:2e⁻ ratio), 70
Porphyrins, 26, 45
Potassium ions, 93, 107, 108
Protohaem, 55, 56
Protonmotive force (p.m.f.), 75, 76, 82,
  95
Proton translocation, 74—6, 80, 81
  stoichiometry, 77—9
Protoporphyrin IX, 45, 54, 55
  structure, 55
Protoporphyrinogenase, 45
Purine nucleotide cycle, 31, 38

Pyridoxal phosphate, 29
Pyruvate, 23—36, 61, 65, 69, 92
  transporting system, 96, 99, 100
Pyruvate carboxylase, 26, 27, 29, 31, 42
Pyruvate decarboxylase, 33, 35
Pyruvate dehydrogenase, 23, 34, 41
  complex, 24, 31—6
  regulation, 36, 42
Pyruvate dehydrogenase kinase, 36, 42
Pyruvate dehydrogenase phosphatase,
  36, 42
Pyruvate kinase, 26
Pyruvate-malate cycle, 30, 31
Pyruvate transporter, 27, 31

Rana pipiens, 117
Red blood cells, 3, 18
Redox potential, 47, 59, 60, 61, 73
Reducing equivalents, 102—5
Respiratory control, 41, 69
Restriction endonuclease, 133, 134
Reversed electron transport, 87, 88
Riboflavine (Vitamin B₂), 50
RNA polymerase, 120
Rotenone, 61, 66, 69, 70, 92, 95, 107
  structure, 66
Rutamycin, 83, 86
Ruthenium red, 97, 108, 109

Saccharomyces cerevisiae, 43, 113,
  115—18, 122, 126, 132, 133
  life cycle, 122, 123
  mitochondrial genetic map, 134
  mitochondrial structure, 2, 3
  petite mutants, 3, 122—6, 128, 131
Schizosaccharomyces pombe, 117
Sea urchins, 3
Septa, 7
Serum albumin, 38, 91
Sodium ions, 107, 109
Sodium pump, 18
Spiramycin, 127
State III respiration, 61, 69, 70
State IV respiration, 61, 69, 70
Steroids, 26, 27
  steroid hormones, 45
Submitochondrial particles, 63—5, 81,
  85, 130
Substrate level phosphorylation, 71, 72
Succinate, 24—6, 30, 47, 50, 60—2, 65,
  70, 75, 76
Succinate dehydrogenase, 23, 24, 32, 50,
  51, 60, 64, 65, 70, 73, 128
Succinate dehydrogenase complex
  (Complex II), 61, 62, 88, 132

Succinic thiokinase, 23, 24
Succinyl CoA, 23–6, 36, 42
Superoxide dismutase, 16
Suppressivity, 125

Terpenoids, 26, 27
Tetrahymena pyriformis, 117
Thiamine pyrophosphate (TPP), 32–5
  structure, 32
Transamination, 29, 36, 38, 103
Transporting systems
  adenine nucleotide, 96, 98, 99, 105,
    106, 110
  antiporter, 95
  aspartate, 97, 101, 103, 104
  biological importance, 102
  calcium ions, 97, 108, 110
  carnitine, 97, 106, 107
  citrulline, 97, 105
  dicarboxylate, 93, 96, 100
  electrogenic, 95, 104, 105
  glutamate, 94, 95, 97, 101, 104, 110
  glutamine, 97, 98, 105, 106
  group translocase, 106, 107
  isolation of, 109, 110
  ornithine, 97, 105
  oxoglutarate, 97, 100, 101, 103
  phosphate, 93, 96, 98, 100, 102
  pyruvate, 96, 99, 100
  tricarboxylate, 93–6, 100–2
  uniporter, 95
Tributyltin chloride, 86
Tricarboxylate transporter, 27, 28, 30,
    31, 93–6, 100–2
Tricarboxylic acid cycle, 17, 22–4, 40,
    47, 65, 75, 132
  ATP yield, 23
  control, 42, 43
  degradative pathways leading to, 25
  efficiency, 23, 24
  gluconeogenesis and, 26, 27

location, 16
reactions of, 24,
role, 25–7
source of biosynthetic precursors, 26
Triethyltin, 127
Tripartite repeating units, 7, 8, 13–15,
    63, 64, 81, 130
  dimensions, 84
  location of ATP synthetase complex,
    13
  structure, 84, 86
Trypanosomes, 3

Ubiquinone (Coenzyme Q), 51–3, 62,
    66, 70, 72, 73–5, 77, 78, 126
  absorption spectrum, 54
  cycle (Q-cycle), 77, 78
  oxidation-reduction, 52, 53
  redox potential, 60
  structure, 53
Uncouplers, 99, 107
Urea, 28, 36–8, 43
Urea cycle, 36–8, 43, 92, 102
  ATP yield, 38
  location, 16, 36, 37
  reactions of, 37
Uridine triphosphate (UTP), 20

Valinomycin, 93, 94, 107
  structure, 108
Venturicidin, 127

Warburg manometer, 68

Xenopus laevis, 117

Yeast, 41, 56, 129, 130
  glucose repression, 131
  see also Saccharomyces cerevisiae,
    Kluyveromyces lactis and
    Schizosaccharomyces pombe